ELECTRONIC DEVICES AND CIRCUIT DESIGN

Challenges and Applications in the Internet of Things

ELECTRONIC DEVICES AND CIRCUIT DESIGN

Challenges and Applications in the Internet of Things

Edited by
Suman Lata Tripathi, PhD
Smrity Dwivedi, PhD

First edition published 2022

Apple Academic Press Inc.
1265 Goldenrod Circle, NE,
Palm Bay, FL 32905 USA

4164 Lakeshore Road, Burlington,
ON, L7L 1A4 Canada

CRC Press
6000 Broken Sound Parkway NW,
Suite 300, Boca Raton, FL 33487-2742 USA

2 Park Square, Milton Park,
Abingdon, Oxon, OX14 4RN UK

© 2022 by Apple Academic Press, Inc.

Apple Academic Press exclusively co-publishes with CRC Press, an imprint of Taylor & Francis Group, LLC

Reasonable efforts have been made to publish reliable data and information, but the authors, editors, and publisher cannot assume responsibility for the validity of all materials or the consequences of their use. The authors, editors, and publishers have attempted to trace the copyright holders of all material reproduced in this publication and apologize to copyright holders if permission to publish in this form has not been obtained. If any copyright material has not been acknowledged, please write and let us know so we may rectify in any future reprint.

Except as permitted under U.S. Copyright Law, no part of this book may be reprinted, reproduced, transmitted, or utilized in any form by any electronic, mechanical, or other means, now known or hereafter invented, including photocopying, microfilming, and recording, or in any information storage or retrieval system, without written permission from the publishers.

For permission to photocopy or use material electronically from this work, access www.copyright.com or contact the Copyright Clearance Center, Inc. (CCC), 222 Rosewood Drive, Danvers, MA 01923, 978-750-8400. For works that are not available on CCC please contact mpkbookspermissions@tandf.co.uk

Trademark notice: Product or corporate names may be trademarks or registered trademarks and are used only for identification and explanation without intent to infringe.

Library and Archives Canada Cataloguing in Publication

Title: Electronic devices and circuit design : challenges and applications in the Internet of things / edited by Suman Lata Tripathi, PhD, Smrity Dwivedi, PhD.
Names: Tripathi, Suman Lata, editor. | Dwivedi, Smrity, editor.
Description: Includes bibliographical references and index.
Identifiers: Canadiana (print) 20210242671 | Canadiana (ebook) 20210242728 | ISBN 9781771889933 (hardcover) | ISBN 9781774639290 (softcover) | ISBN 9781003145776 (ebook)
Subjects: LCSH: Electronic apparatus and appliances. | LCSH: Electric circuits. | LCSH: Electronics—Materials. | LCSH: Internet of things.
Classification: LCC TK7870 .E44 2022 | DDC 621.3—dc23

Library of Congress Cataloging-in-Publication Data

Names: Tripathi, Suman Lata, editor. | Dwivedi, Smrity, editor.
Title: Electronic devices and circuit design : challenges and applications in the internet of things / edited by Suman Lata Tripathi, PhD, Smrity Dwivedi, PhD.
Description: First edition. | Palm Bay, FL, USA : Apple Academic Press, [2022] | Includes bibliographical references and index. | Summary: "This new volume, Electronic Devices and Circuit Design: Challenges and Applications in the Internet of Things, offers a broad view of the challenges of electronic devices and circuits for IoT applications in a concise way. The book presents the basic concepts and fundamentals behind new low power, high-speed efficient devices, circuits, and systems with new technology in addition to CMOS. It aims to help develop an understanding of new materials to improve device performance with smaller dimensions and lower costs. It looks at the new methodologies to enhance system performance and provides key parameters to explore the devices and circuit performance based on smart applications. The volume bridges the gap for researchers working on different areas of smart devices, circuits, and systems with IoT applications. The fundamental principles of electronic device and circuit design are discussed in a clear and detailed manner along with explanatory diagrams. Various IoT-based smart applications in different fields along with challenges and issues along with prospects of IoT for future applications have also been discussed. The chapters delve into myriad aspects of circuit design, including MOSFET structures depending on their low power applications for IoT-enabled systems, advanced sensor design and fabrication using MEMS, indirect bootstrap techniques, efficient CMOS comparators, various encryption-decryption algorithms, IoT video forensics applications, microstrip patch antennas in embedded IoT applications, real-time object detection using sound, IOT and nanotechnologies based wireless sensors, and much more. It also introduces logic BIST using BS-CFSR for electronic devices that can be part of many applications in many industries, such as defense, automotive, banking, computer, healthcare, networking, and telecommunication. This book provides a treasure house of up-to-date information on many aspects of electronic devices and circuit design, making it highly beneficial for researchers, scientists, faculty, students, and industry professionals"-- Provided by publisher.
Identifiers: LCCN 2021028094 (print) | LCCN 2021028095 (ebook) | ISBN 9781771889933 (hbk) | ISBN 9781774639290 (pbk) | ISBN 9781003145776 (ebk)
Subjects: LCSH: Internet of things--Equipment and supplies. | Electronic apparatus and appliances.
Classification: LCC TK5105.8857 .E43 2022 (print) | LCC TK5105.8857 (ebook) | DDC 004.67/8--dc23
LC record available at https://lccn.loc.gov/2021028094
LC ebook record available at https://lccn.loc.gov/2021028095

ISBN: 978-1-77188-993-3 (hbk)
ISBN: 978-1-77463-929-0 (pbk)
ISBN: 978-1-00314-577-6 (ebk)

About the Editors

Dr. Suman Lata Tripathi
FIETE, SMIEEE, LMISC
MIEEE EDS, PES, CN, IAS
Professor, VLSI Design, School of Electronics and Electrical Engineering, Lovely Professional University, Phagwara 144411, Punjab, India. Email: suman.21067@lpu.co.in

Suman Lata Tripathi, PhD, is associated with Lovely Professional University as a Professor with more than 17 years of experience in academics. She has published more than 45 research papers in refereed journals and conferences and holds several patents. She has organized workshops, summer internships, and expert lectures for students. She has worked as a session chair, conference steering committee member, editorial board member, and reviewer in international and national IEEE journals and conferences. She has been awarded a Research Excellence Award in 2019 at Lovely Professional University and has received the best paper award at IEEE ICICS–2018. She is also associated as an editor/author of books and book series with publishers such as CRC Press, Taylor & Francis, Springer, Nova Science Publishers, Scrivener Publishing Wiley, IGI Global Publisher, Apple Academic Press, etc. Her area of expertise includes microelectronics device modeling and characterization, low power VLSI circuit design, VLSI design of testing and advance FET design for IoT and biomedical applications, etc. She also has an interest in the area of the energy-efficient systems, green energy, and renewable energy. She has completed her PhD in the area of microelectronics and VLSI from MNNIT, Allahabad, India; her MTech in Electronics Engineering from UP Technical University, Lucknow; and BTech in Electrical Engineering from Purvanchal University, Jaunpur, India.

Dr. Smrity Dwivedi
SMIEEE, LMVEDA, MIEEEPS
Assistant Professor (Senior Grade), IIT BHU, Varanasi 221005, UP, India. Email: sdwivedi. ece@iitbhu.ac.in

Smrity Dwivedi, PhD, is currently working as an Assistant Professor in the Department of Electronics Engineering at the Indian Institute of Technology, Banaras Hindu University (IITBHU), Varanasi, India. She is an IEEE senior member and an active member of IEEE plasma science, IEEE antenna, and wave propagation sections, and a life member of VEDA (Vacuum Electron Devices and Applications). She is also a reviewer of many international and national journals. She has published many papers in prestigious international and national journals as well as in many international and national conferences. Her current areas of specialization are high-power microwave tubes, smart antenna design using the newest technologies, MMICs, etc. Currently, she is guiding MTech and PhD scholars in the area of antenna and microwave engineering. She has completed her PhD from IIT BHU, Varanasi, India, in high-power microwave tubes.

Contents

Contributors .... *ix*
Abbreviations .... *xiii*
Preface .... *xvii*

1. **An Overview of Future IoT Systems: Applications, Challenges, and Future Trends** 1
   Meet Kumari

2. **Design and Analysis of Advanced MOSFET Structures for IoT Applications** 19
   Suman Lata Tripathi

3. **Integration of MEMS Sensors for Advanced IoT Applications** 33
   Anuj Kumar Goel

4. **CMOS Bootstrap Driver** 51
   Abhishek Kumar

5. **Implementation of Low-Power BIST Using Bit Swapping Complete Feedback Shift Register (BSCFSR)** 69
   Ravi Trivedi and Sandeep Dhariwal

6. **A Review of a Low-Power CMOS Comparator** 79
   Tejender Singh and Suman Lata Tripathi

7. **Encryption and Decryption Algorithms for IoT Device Communication** 97
   Ananya Dastidar and Sonali Mishra

8. **Future of Video Forensics in IoT** 113
   Sunpreet Kaur Nanda and Deepika Ghai

9. **Role of Microstrip Patch Antenna for Embedded IoT Applications** 135
   Amandeep Kaur, Praveen Kumar Malik, and Ravi Shankar

10. **Sensible Vision Using Image Processing** 157
    Kanwaljeet Singh, Shivang Tyagi, and Baljeet Kaur

11. **Detection of Blocking Artifacts in JPEG Compressed Images at Low Bit Rate for IoT Applications**.................................. 173
    Anudeep Gandam, Jagroop Singh Sidhu, and Manwinder Singh

12. **Quality of Service Provisioning in Mobile Ad Hoc Networks**............. 197
    Manwinder Singh and Kamal Kumar Sharma

13. **Solution of Automatic Generation Control of Multi-Area Power Plant Strategy with a Nonconventional Energy Source in Cooperation with Smart Controllers**.................................. 209
    Krishan Arora and Tarun Dhandhel

14. **IoT-Based Technology for Smart Farming**... 223
    Vikalp Joshi and Manoj Singh Adhikari

15. **Orthogonal Frequency Division Multiplexing for IoT**........................ 243
    Arvind Kumar and Rajoo Pandey

16. **Fading Channel Capacity of Cognitive Radio Networks**..................... 269
    Indu Bala

17. **Touch Screen Mobile Phones Form Analysis Using Kansei Engineering**.. 279
    Vivek Sharma and Kamalpreet Sandhu

Index.. *291*

Contributors

Krishan Arora
School of Electronics and Electrical Engineering, Lovely Professional University, Punjab.
E-mail: Krishan.12252@lpu.co.in

Indu Bala
School of Electronics and Electrical Engineering, Lovely Professional University, Punjab, India.
E-mail: i.rana80@gmail.com; indu.23298@lpu.co.in

Tarun Dhandhel
School of Electronics and Electrical Engineering, Lovely Professional University, Punjab

Ananya Dastidar
Department of Instrumentation and Electronics Engineering, College of Engineering and Technology, Bhubaneswar, Odisha, India. E-mail: adastidar@cet.edu.in

Sandeep Dhariwal
Department Of ECE, Alliance College of Engineering and Design, Alliance University, Bengaluru 562106, Karnataka, India. E-mail: dhariwal.vlsi@gmail.com

Anudeep Gandam
Department of Electronic and Communication Engineering, IKG-Punjab Technical University, Jalandhar, Punjab, India. E-mail: gandam.anu@gmail.com

Deepika Ghai
Assistant Professor, School of Electronics and Electrical Engineering, Lovely Professional University, Punjab

Anuj Kumar Goel
Department of Electronics and Communication Engineering, Chandigarh University, Punjab, India.
E-mail: anuj40b@gmail.com

Himani Jerath
School of Electronics and Electrical Engineering, Lovely Professional University, Punjab, India.
E-mail: jerathhimani29@gmail.com

Amandeep Kaur
Department of Electronics and Communication Engineering, Lovely Professional University, Jalandhar, Punjab, India. E-mail: aman.dhaliwal18@gmail.com

Baljeet Kaur
GNE College of Engineering, Ludhiana, India. E-mail: baljeetkaur@gndec.ac.in

Meet Kumari
Department of Electronics and Communication Engineering, Chandigarh University, Mohali, Punjab, India. E-mail: meetkumari08@yahoo.in

Abhishek Kumar
School of Electronics and Electrical Engineering, Lovely Professional University, Punjab, India.
E-mail: abhishek.15393@lpu.co.in

Arvind Kumar
Department of Electronics & Communication Engineering, NIT, Kurukshetra, India.
E-mail: arvind_sharma@nitkkr.ac.in

Praveen Kumar Malik
Department of Electronics and Communication Engineering, Lovely Professional University, Jalandhar, Punjab, India

Sonali Mishra
Department of Instrumentation and Electronics Engineering, College of Engineering and Technology, Bhubaneswar, Odisha, India

Sunpreet Kaur Nanda
Research scholar, School of Electronics and Electrical Engineering, Lovely Professional University, Punjab
Assistant Professor, EnTC Department, P. R. Pote College of Engineering and Management, Amravati

Rajoo Pandey
Department of Electronics & Communication Engineering, NIT, Kurukshetra, India

P. Raja
School of Electronics and Electrical Engineering, Lovely Professional University, Punjab, India.
E-mail: raja.21019@lpu.co.in

Kamalpreet Sandhu
Department of Product and Industrial Design, Lovely professional University, Phagwara

Ravi Shankar
Department of Electronics and Communication Engg., Lovely Professional University, Jalandhar, Punjab

Vivek Sharma
Department of Product and Industrial Design, Lovely Professional University, Phagwara.
E-mail: vivek.24810@lpu.co.in

Kamal Kumar Sharma
School of Electronics and Electrical Engineering, Lovely Professional University, Punjab, India

Dushyant Kumar Singh
School of Electronics and Electrical Engineering, Lovely Professional University, Punjab, India.
E-mail: Dushyantkumarsingh83@lpu.co.in

Kanwaljeet Singh
School of Electronics and Electrical Engineering, Lovely Professional University, Phagwara, India.
E-mail: Kamal1997@gmail.com

Tejender Singh
School of Electronics and Electrical Engineering, Lovely Professional University, Punjab
CMR Institute of technology, Hyderabad, Telangana, India. E-mail: tejendersingh27@gmail.com

Manwinder Singh
Department of Electronic and Communication Engineering, Lovely Professional University Phagwara Jalandhar, Punjab, India. E-mail: manwinder.25231@lpu.co.in

Contributors xi

Suman Lata Tripathi
School of Electronics and Electrical Engineering, Lovely Professional University, Punjab, India.
E-mail: tri.suman78@gmail.com

Ravi Trivedi
Physical design engineer, Digicomm Semiconductors Pvt. Ltd., Bengaluru 560103, Karnataka, India.
E-mail: ravi.trivedi221192@gmail.com

Shivang Tyagi
School of Electronics and Electrical Engineering, Lovely Professional University, Phagwara, India.
E-mail: Shivang.tyagi90@gmail.com

Abbreviations

3GPP	3rd Generation Partnership Project
ABC	Artificial Bee Colony
AC	air conditioner
ACE	area control error
ACSR	adjacent conjugate symbol repetition
ADC	analog–digital converter
AGC	automatic generation control
ANN	artificial neural network
ARTP	adaptive rate and transmission power
ATM	automatic teller machine
ATP	adaptive transmission power
BER	bit error rate
BDCT	block discrete cosine transform
BDL	dynamic bootstrap logic
BOX	buried oxide region
BPP	bits per pixel
BS-CFSR	bit swapping complete feedback shift register
CC	conjugate cancellation
CFO	carrier frequency offset
CIR	carrier-to-interference ratio
CL	carry-look ahead
CP	cyclic prefix
CUT	circuit under test
CWSI	crop water stress index
DCT	discrete cosine transform
DG	double gate
DIBL	drain-induced barrier lowering
DL	downlink
DM FET	dielectrically modulated FET
DTDLC	double-tail current dynamic latch comparator
DTDLC-CLK	two-stage dynamic comparator without inverted lock
DTN	delay-tolerant network
ESs	embedded systems

FAO	Food and Agriculture Organization
FCI	Food Cooperation of India
FFT	fast Fourier transform
FLC	fuzzy logic controller
GA	genetic algorithm
GAA	gate-all-around
GD	gradient descent
ICI	intercarrier interference
IDE	integrated development environment
IDFT	inverse discrete Fourier transform
IoT	internet of things
LFC	load-frequency control
LFSR	linear feedback shift register
MANET	mobile ad hoc network
MPC	model predictive control
MDTDLC	modified double-tail current dynamic latch comparator
NB-IoT	narrowband IoT
NR	new radio
OFDM	orthogonal frequency division multiplexing
OLSR	Optimized Link State Routing protocol
PDR	packet delivery ratio
PI	proportional integral
PID	proportional with integral derivative
PRx	primary user receiver
PSO	particle swarm optimization
PT	pass transistor
PTx	primary user transmitter
PU	primary user
PV	photovoltaic cell
RC	ripple carry
ROM	read only memory
RTOS	real-time operating system
SCEs	short channel effects
SE	scan enable
SNR	signal-to-noise ratio
SOI	silicon on insulator
SRx	secondary user receiver
SS	subthreshold slope

STx	secondary user transmitter
STDLC	single-tail current dynamic latch comparator
SU	secondary user
SWDI	soil moisture is compared to water deficit index
TG	tri-gate
UAV	unmanned aerial vehicles
UL	uplink
VANET	vehicular ad hoc network
WLAN	wireless local area network
WSN	wireless sensor networks

Preface

The objective of this edition is to provide a broad view of electronic device and circuit challenges for IoT application in a concise way for fast and easy understanding. This book provides information regarding almost all the aspects to make it highly beneficial for all the students, researchers, and teachers of this field. Fundamental principles of electronic device and circuit design are discussed herein in a clear and detailed manner with an explanatory diagram wherever necessary. All the chapters are illustrated in simple language to felicitate readability.

Chapter Organization
This book is organized in 17 chapters.

Chapter 1 discusses various IoT-based smart applications in different fields. Also, major open challenges and issues along with prospects of IoT for future applications have also been discussed.

Chapter 2 is mainly focused on the fundamental concept to study and analyze new MOSFET structures depending on their low power applications for IoT-enabled systems.

Chapter 3 deals with advanced sensor design and fabrication using MEMS that are used in IoT systems, as soil moisture sensors, temperature sensors, gyroscopes, accelerometers, biosensors, pressure sensors, magnetometers, optical actuator, gas sensors, etc.

Chapter 4 deals with the indirect bootstrap technique of gate deriver for a large capacitive load at the output when the bootstrap technique at gate output of the device is not impressive in reducing the transition speed of the output.

Chapter 5 introduces logic BIST that can be part of many applications such in the defense, automotive, banking, computer, healthcare, networking, and telecommunication industries. The availability of low power BIST using BS-CFSR for electronic devices is the key to the success of IoT applications.

Chapter 6 is related to different comparators where a comparison is done between different models of comparators. The efficient comparators are also explored for a wide range of Internet of Things applications, such as health care, environmental monitoring, and manufacture unit.

Chapter 7 gives a detailed analysis of designing various CMOS comparators, that reduces the power consumption, uses low voltage and detects the presence of a signal at high speed. The chapter discusses various encryption–decryption algorithms that can be implemented during the communication taking place in an Internet of Things network.

Chapter 8 emphasizes the IoT video forensics applications, and limitations in the currently available video forensics tools as well as the future of this research.

Chapter 9 describes the role of the microstrip patch antenna discussed in embedded IoT applications. To make embedded circuits more compact, smaller sized and lightweight antennas are desired. This tremendous growth in IoT field has placed demand over multiband and ultrawideband antennas.

Chapter 10 explores real-time object detection to alert the user about the surrounding objects and their position by using sound. Sensible vision can give fast response with the help of some algorithms. Sensible vision has a Bluetooth system making a connection between system and any kind of sound system.

Chapter 11 presents a modified model based on the assumption of the large variation of pixel values of adjacent pixels across the block boundary in comparison to pixels away from the boundary is designed. The experimental results show that the proposed method outperforms while detecting blocking artifacts as compared to other post-processing methods.

Chapter 12 deals with OLSR routing protocol along with the ABC algorithm that has been performed in order to minimize the problem related to the routing protocols and to reduce the energy consumption rate of the simulation work. There are certain steps that are required to be followed in order to create a simulated network in MANET.

Chapter 13 describes the automatic generation control of wind, hydro, and thermal power plants that has been connected with progressive controllers. Particle Swarm Optimization is highly productive and reliable in calculations of dissimilar gains in load frequency control.

Chapter 14 deals with the capability of IoT and nanotechnologies-based wireless sensors generally known as mote.

Chapter 15 emphasizes the 5G new radio (NR) technology on diverse services like Internet of Things, multimedia broadcast network (MBN), and enhanced mobile broadband (eMBB).

Chapter 16 provides a concise survey of a communication system in which an unlicensed user access is licensed spectrum opportunistically and regulates its system parameters such as transmission power based on channel sensing information under received power constraints.

Chapter 17 deals with Kansei engineering, a methodology by which user emotions or perception about a product can be identified more precisely and can be linked with product forms elements such as shape, size, and color, etc. In this work, a consumer preference-oriented design method of products' form perception is proposed on a touchscreen mobile phone.

CHAPTER 1

An Overview of Future IoT Systems: Applications, Challenges, and Future Trends

MEET KUMARI*

Department of Electronics and Communication Engineering, Chandigarh University, Mohali, Punjab, India

*E-mail: meetkumari08@yahoo.in

ABSTRACT

The Internet of things (IoT) is an innovative digital world technology that extends to connect several digital devices with various sensing and computing digital devices in the context of smart applications. It is a future-based technology that provides the solutions for full-duplex communication information flow, minimizing energy wastage, increasing energy demand, security, flexibility, and reliability in the traditional interconnected network. IoT helps in connectivity of various smart devices anytime and anywhere for analysis, monitoring, and controlling the future smart applications. It also provides the interconnectivity, automation, and tracking of these devices through the Internet. In this chapter, an overview of the future-based IoT system has been presented. Firstly, the IoT-based layered architecture, its standard bodies and requirements along with summarized review for smart applications has been discussed. Secondly, various IoT-based smart applications in different fields have been mentioned. Also, major open challenges and issues along with prospects of IoT for future applications have also been discussed.

1.1 INTRODUCTION

The Internet of things (IoT) and telecommunication field are defined as the network of physical devices, objects, vehicles, infrastructure, and other things, embedded with hardware, software, network, and sensor devices to collect and interchange the information. IoT is a current and future technology that carries huge change in business sector.[1] It allows for smart work by creating human-to-machine and machine-to-machine interactions.[2]

In addition, the primary strength of the IoT technology is its large impact on various aspects of daily life and users' behaviors and it can be seen in both private and public working fields of users. Some of the potential applications of IoT are e-health, e-learning, assisted living, smart vehicles, automation, industrial, domestic, manufacturing, and process management.[1]

IoT was first described in 1999 and follows the approach of Internet-based applications in 1990. In IoT, usually about 100 billion devices will be interconnected to the Internet by 2020. Figure 1.1 shows the predicted growth in the population versus Internet-based smart devices from 2015 to 2025. It is clear from the figure that as the world population increases to 7.99 billion, the number of Internet-based smart devices will be 75.44 billion. Thus, smart devices will increase by 2–5 times of world population from 2015 to 2025, respectively.[3]

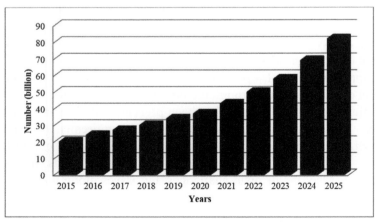

FIGURE 1.1 Comparative analysis of estimated world population and projected number of IoT-based smart devices.[8]

The smart world is a compound ecosystem featured by the major use of information and communication technologies to make the world more sustainable, attractive, innovative, and entrepreneurship. The prime stakeholders consist of application developers, citizens, service providers, government, research community, public service providers and platform developers. In context, IoT system plays a significant and fundamental role in large-scale distributed network development.[6] Further, the IoT applications can be basically categorized as network type, flexibility, repeatability, scalability, and users involvement.[4]

In this chapter, the future-based IoT system has been overviewed. In Section 1.2, the IoT-based layered architecture, its standard bodies and requirements, has been presented. In Section 1.3, summarized review of IoT system for various applications has been discussed. In Section 1.4, IoT-based smart applications in different fields have been illustrated. In Sections 1.5 and 1.6, IoT-based open challenges along with issues and future prospective of IoT have been discussed, respectively. Finally, a conclusion is drawn in Section 1.7.

1.2 ARCHITECTURE OF IOT SYSTEM

The basic architecture of IoT system can be understood by its layered architecture, requirements, and standard bodies. These are described in the following sections.

1.2.1 LAYERED STRUCTURE OF IOT SYSTEM

The basic IoT system based on smart applications consists of five layers. The prime layers of IoT based on smart applications are shown in Figure 1.2[3,5]:

a. **Device layer:** It is also known as perception or object or sensing or edge layer. It is responsible for hardware or physical connection of devices in IoT. It also collects the data, exchanges the information, and reads the data in IoT applications.[3,5]

b. **Network layer:** It joins all the IoT devices with each other to interchange the data. Here, the data is transmitted from source to destination via key methods such as Wi-FI, ZigBee, 3G,

Bluetooth, visible light communication (VLC), and infrared technology.[3,5]

c. **Middleware layer:** It combines the IoT-based services to requester by using name and address. It permits the IoT application services to work in distributed environment. Again, it collects the information from network and stores it in the database.[3,5]

d. **Application layer:** It enables high-quality services to meet the users' requests. It also provides the information that evaluates and integrates the received information from the abovementioned layers.[3,5]

e. **Business layer:** It manages the whole activities, services flow-charts, and graphs. It is responsible for the achievements of IoT-based services in business model.[3,5]

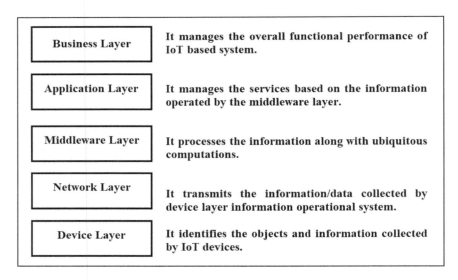

FIGURE 1.2 Layered architecture of IoT-based services in smart applications.[8]

1.2.2 MAJOR STANDARD BODIES OF IOT SYSTEM

Various major standard bodies for enabling smart applications in IoT are shown in Figure 1.3[4]:

An Overview of Future IoT Systems

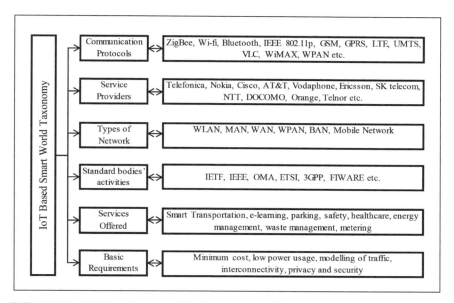

FIGURE 1.3 IoT-based smart applications taxonomy.[6]

1.2.3 BASIC REQUIREMENTS FOR IOT SYSTEM

The basic requirements for implementing IoT-based smart applications are shown in Table 1.1.[6]

1.3 LITERATURE SURVEY

The literature review of IoT system in different applications is summarized in Table 1.2.

From this summarized literature review, it is clear that IoT is the basis of current and incoming future smart applications due to the increasing usability of users day by day. However, some have issues and challenges. Hence, many researchers are working on IoT-based system to make them real for users.

TABLE 1.1 Requirements of IoT System.[4]

Sl. No.	Technology	Frequency	Bit rate (b/s)	Range (m)	Latency (ms)	Power	Applications
1	ZigBee	2.4 GHz, 868/915 MHz	250 M	100	16	Less	Metering, e-healthcare
2	Bluetooth	2.4 GHz	25 M	10	100	Less	e-healthcare
3	Wi-Fi	2.4 GHz	54/6.75 G	140	46	Average	Waste and energy management
4	IEEE 802.11p	5.85 GHz	6 M	100	0.5	Less	Transpiration
5	DSRC	5.8/5.9 GHz	6 M	1000	0.2	Less	Transpiration
6	DASH7	433/915 MHz	55.5/200 K	1000	15	Less	Transpiration and automation
7	GSM	850/900/1800/1900 MHz	80–348 K	30000	3000	High	Metering, e-healthcare, and transpiration

1.4 APPLICATIONS

Various smart world applications of IoT-based system are described in the following sections.

TABLE 1.2 Summarized Literature Survey of IoT System in Different Fields.

Sl. no.	Author (date of publication)	Remarks
1	D. Mourtzis et al. (2016)	The paper shows that the adoption of IoT in manufacturing provides the modern, digital, cost–effective, etc. manufacturing system. However, it also increases the manufacturing volume data, but the presence of IoT cyber-production paradigm enables for flexible production.[6]
2	Mandrita Banerjee et al. (2018)	The paper shows that secure IoT data can be exchanged between IoT database and systems through blockchain technology approaches. The optimization of blockchain platforms, for example, open-source Coco Framework reduces the energy usage and improves the security with efficient services.[7]
3	M. Aabid A Majeed et. al. (2017)	The paper illustrates that the radio-frequency identification (RFID) technology adopted by supply chain manufactures must be flexible and affordable in digital world and Industry 4.0 to reconstruct the IT sector in the market at best services and consumer shopping experience.[8]
4	Julien Mineraud et al. (2016)	In this paper, various IoT platforms have been discussed to evaluate the gap analysis for supporting distributed hardware integration with sufficient data management methods to support the ecosystem at dedicated marketplace for IoT.[9]
5	Nikos Bizanis et al. (2016)	In this paper, the fusion between virtualized SDN network and IoT has been examined. It shows that the integration of SDN NV is an effective way for forthcoming IoT-based 5G networks.[10]
6	Sravani Challa et al. (2017)	The paper shows that IoT environment can be secured by signature-based authenticated proposed key scheme such as Burrows–Abadi–Needham logic. It shows that this approach defiantly reduces the cost compared to other existing approaches.[11]

1.4.1 SMART AGRICULTURE

The smart IoT-based agriculture turns on and off the IoT devices automatically in real time on the basis on sensed information such as CO_2 concentration, greenhouse effect, soil condition, and humidity. Mostly wireless sensor networks (WSNs) are used for agriculture system for sensing the environmental parameters such as energy consumption, topology construction, and cyber-attack security. Firstly, these parameters are captured by different multihop and self-organizing sensors. Then these sensed data transmitted to cloud server through cellular network (e.g., 3G/4G/5G). At this stage, the smart IoT-based agriculture is ready for processing automatically anytime to monitor ecological information for ensuring intelligent management and suitable growth of crops. For real-time monitoring, mobile communication system, and remote areas, the agriculture-based mobile phone apps in mobile phone or personal computers are used. This communication helps in modifying the crops' needs, warning messages, etc. to respective farmers. Further, it can be applied in supply chain management in agriculture for ensuring food safety, quality, and production process.[12]

1.4.2 SMART INDUSTRY

Various smart objects have been employed in industrial production applications such as purchasing raw material, sale, and inventory enterprises to enhance the supply chain efficiency with less cost in supply chain management system. It provides the sensing as well as decision-making capability for enterprises in process optimization of production process and material consumption monitoring. Also, it shows the industrial model for production and maintenance management in manufacturing operation. It acquires the real-time data for energy consumption. In motoring process, it focuses on energy management equipment monitoring system. Some of the smart industry applications are remote monitoring of oil, gas, mines, etc. stations for ensuring safety and real-time monitoring of polluted sources, saving resources and energy with accurate identification, response, and effective control. Industry 4.0 is the foundation for smart industry. It integrates the industrial technology to create an energy-efficient and adaptive intelligent factory. Its primary aim is to define the features of sustainable developments based on smart paradigm plant management of industry with expected benefits.[12]

1.4.3 SMART HOME

Smart home refers to an innovative and flexible home management system utilizing several technologies like network communication, automatic control, security defense, and audio/video. It enhances the home living comfort, safety, artistry, convenience, and friendly environment. Here, the smart objects or devices are interconnected with one another through Wi-Fi networks for long- and short-range communications. These devices can be used for remote control smart home system. Some of the examples of smart home are sound awareness technology for detecting users' behavior, fault tolerance, safety, quality, etc. to handle the complex changes in the home environment. This front-end smart home device must be scalable, reliable, cost-effective, power saver, and versatile in standardization. As compared to normal home, the smart home consists of traditional living features with full range of interconnectivity among appliances, equipment, and buildings to provide the best comfort for family.[12] Figure 1.4 shows the summarized possible application of IoT in the current world.

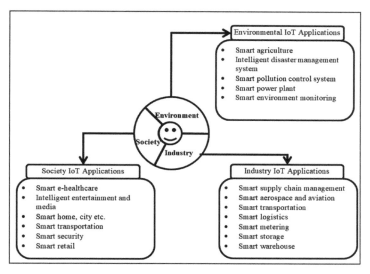

FIGURE 1.4 IoT-based smart world applications.[9]

1.4.4 SMART TRANSPORTATION SYSTEM

Intelligent and smart transportation is the future of vehicular system. In this system, the mobile nodes can connect and disconnect automatically and freely on Internet of vehicular system. For this, infrared, VLC, RFID, and Bluetooth are used for rigid and versatile devices to interchange the information in transportation network. This helps in cloud-based and low-power services by collecting driver behavior and travel information through electronic sensor system. The possible technologies for smart transportation system are vehicular ad hoc network (VANET), TOME, etc. These technologies facilitate data collection, data processing, data exchanging, and data analysis via computers or mobile phones at accurate time and place.[12]

1.4.5 SMART HEALTHCARE SYSTEM

It presents various health-related fields such as patient's disease symptom monitoring for guidance and rehabilitation monitoring for patients' daily activity and in surgery rooms. Smart healthcare has encouraged the development of smart wearable mobile devices. Some of the smart wearable mobile devices are smartphones, smart watches, etc. to monitor the patient's heart rate, sleep sate, blood pressure, and other activities. This opens up new directions in medical filed by sensing data with wearable devices for analyzing patients' health, behavior, and suggestions for further improving the health. It also helps in remote and real-time checking of patients by doctors. Here, the wearable smart devices must be convenience, cost-effective, safer, and easily usable by the user. In future, it can be further enhanced by cloud computing, big data, and deep neural networks.[12]

1.4.6 INTELLIGENT METERING

This is one of the smart IoT applications in smart grids, water flows, and calculations of soils stoke to generate the huge amount of data from several sources. Here, the long-term processing is taken on dedicated as well as powerful machines. A smart meter helps in electronically recording the electric energy consumption between control and meter system. After collecting and analyzing the record, smart meter data in IoT system allows for decision-making in predicting electric energy consumption. Further, it also analyzes the crises and fulfills the objectives and demands with minimum cost.[13]

1.4.7 SMART GRID

It is a new innovative power grid to manage and distribute electricity among suppliers and consumers. It is a full-duplex communication and computing technology to enhance real-time safety, efficiency, reliability, scalability, controllability, and monitoring. As integrating and decentralizing the renewable energy are the major challenges in energy system, smart grid in electricity system helps to manage the volatile effect of distributed energy resources by following government laws and regulations. In smart grid systems, various grid sensors and objects rapidly and continuously generate the control loop and protection data. Then the human-to-machine and machine-to-machine interactions to control the system commends processed to fulfill the reporting and visualizing requirements.[13]

1.4.8 E-COMMERCE

IoT offers the well-designed application tools to process the decision-making real-time smart e-commerce applications. It consists of distributed system having high volume and data processing characteristics. These result in increase in revenue growth, customer size, accuracy, product optimization, risk management, and enhanced segmentation.[13]

1.4.9 SMART CITIES

IoT information generated from smart cities provides innovative opportunities with high-efficiency gain through a systematic analytical infrastructure to analyze the data. Several IoT devices collect and exchange information through smart Internet environment. Further, the presence of big data and cloud computing reduces the cost of system. Thus, the role of IoT in smart cities can significantly improve each sector of economy of country.[13]

1.4.10 SMART ENVIRONMENT

IoT has the potential to improve the quality of life of human by enhancing the environment through improving the air quality and minimizing the pollution. For this, stations are developed and set across the city to collect data using the sensors on vehicle wheels such as car and bike. Then the floating car data helps in estimating the unknown air quality through global positioning system (GPS), Wi-Fi, smartphones, and sensors. After decision-making through machine and deep learning, where real-time data flow, road networks, human mobility, etc. are integrated for processing using artificial neural networks. Further, IoT can be used for monitoring environment conditions on the basis of spatial as well as temporal maps for detecting the natural disaster.[14]

1.4.11 SMART GRID

IoT provides the observation of energy consumption as well as motoring greenhouse gasses emission across temporal scale for efficient energy usage. For this, smart grid gathers the big data to enhance the performance of power system. It also enables to collect an act on real-time data for power consumption and generation to enhance the efficiency, reliability, etc. of power distribution system.[14]

1.4.12 SMART ENERGY

IoT can support big data decision-making to power supply with increasing demand of citizens and others expectations. It can be done by real-time energy and data to respond to market fluctuations. This increases the decision-making capability to reduce or enhance the production in the presence of energy cost and energy stress. It also optimizes and controls the energy integration, consumption, and data sensing in energy-consuming devices with cost-effectiveness IoT system.[14]

1.4.13 SMART WASTE MANAGEMENT

E-waste is the fastest-growing waste management system. For greening, this field through technologies reduces the pollutants from environment for recycling waste and reusing. It leads to reducing the environmental hazard, energy consumption, and resources and improving economic development.[15]

1.5 OPEN ISSUES AND CHALLENGES

Although IoT has various advantages that enable to solve different problems in different sectors, it has various open issues and challenges as discussed in the following sections.[12]

1.5.1 PRIVACY

It is one of the primary challenges in IoT system. It arises because when system wants to restore the sensitive information using analytical tools such as big data although the anonymous users generated that information. It should be secure and protected from anonymous interference by data privacy encryptions of IoT data. Moreover, the presence of homogeneous data types, devices, and nature to generate the data and communication protocols leads to unsure IoT data. Thus, IoT system should be authenticated by assigning a unique identification to each IoT device. This will enhance eth security and privacy in IoT system.[13]

1.5.2 DATA MINING

It provides efficient and best descriptive or predictive methods for IoT system for the generalization of new data. But the presence of cloud computing and big data in IoT system generates the data extraction and exploration challenges due to high volume, variety, and velocity. For this, various programming models and algorithms are proposed.[13]

1.5.3 VISUALIZATION

It is again a primary challenge in IoT system. The presence of large size and high-dimensional data in IoT leads to difficulty in data virtualizations. It is a challenging task in distributed and diverse data in IoT system. For this, various virtualization algorithms can be used for independent task in IoT with high scalability and high functionality and less response time.[13]

1.5.4 INTEGRATION

It provides the different formats of single data from different sources as well as stored data. The generated data can be subdivided into database that consists of columns or rows; semistructured data, for example, JSON and HTML files; and unstructured data, for example, text, image, audio, and video. It is very tough to integrate the different data formats and it should be solved in future IoT system. Further, the adjustment of structured, semistructured, and instructed data is another challenge in IoT data integration.[13]

1.5.5 MULTIPLE NETWORK AND LOCATIONS

As the data stored in multiple locations in IoT system that may have different jurisdictions, the storage of such data can raise the series problem in IoT system. Thus, some standard techniques may be used to analyze and examine the multiple network and locations issues.[13] Major open challenges and issues in IoT system are given in Table 1.3.

TABLE 1.3 Open Issues and Challenges in IoT.[2]

Sl. no.	Issues	Open challenges
1	Security	Design, security and cost trade-offs, metrics, standards, regulations, obsolescence of devices, authentication, confidentiality, controllability, upgradability.
2	Privacy	Data collection fairness, usability fairness, transparency, enforcement, privacy expectations, design privacy, identification.
4	Interoperability	User desire, cost constraint, technical risk, legal system, schedule risk, conformation.
5	IoT standards	Growth of standards efforts.
6	IoT legal, rights, and regulatory	Protection of data, data discrimination, liability of devices, public safety.
7	Economic	Investment.
8	Development	Resources in infrastructures.
9	Networking	Background noise, environmental effects.

1.6 FUTURE SCOPE

As IoT is the future technology, in spite of some major challenges, it is motivated to use in future applications. Some of the major future scopes of IoT system are as follows:

a. **Blockchain in IoT system:** Blockchain technology is the solution of privacy and security challenges for future IoT system by eliminating the core server concept of IoT to allow the data flow through blockchain database with authentication. It has lots of application in IoT technology such as agriculture, energy, food, finance, healthcare, manufacturing, transport, business, and distribution.[2]
b. **WSNs in IoT System:** It allows for various logical decision-making methods to design the IoT-based system for WSNs. But it also has challenge in partition of system traffic in logical phases that affect the WSN functionality.
c. **Cloud computing in IoT system:** It can correctly handle the data in IoT-based system. It is powerful computing analytical capacity system having storage capability in emergency situations as well as decision-making capability.[12]

1.7 CONCLUSION

IoT is expected as the future-based technology due to its various future-based smart applications. In this chapter, IoT-based applications, their challenges, and future scope have been discussed. The presence of wide and open interaction of IoT devices helps in connecting various smart devices to send and receive their data with one another. However, the data sharing can be exploited by intruders depend on the wireless technology used for IoT system. Here, the basic key requirement for enabling smart applications in IoT environment has been enumerated. Furthermore, open research challenges for IoT-based applications as future research opportunities have been discussed. It is concluded that IoT systems must be enabled to incorporate the smart IoT world solutions for safe, flexible, user-friendly, secure and cost-effective IoT environment by using future-based technologies such as blockchain, WSN, and cloud computing.

Otherwise, the users may lose their interest and trust in future-based IoT applications.

KEYWORDS

- **Internet of things**
- **smart applications**
- **IoT architecture**
- **challenges in IoT**

REFERENCES

1. Stergiou, C.; Psannis, K. E.; Kim, B.; Gupta, B. Secure Integration of Internet-of-Things and Cloud Computing Secure Integration of IoT and Cloud Computing. *Future Gener. Comput. Syst.* **2018**, *78*, 964–975; https://doi.org/10.1016/j.future.2016.11.031.
2. Manoj, N.; Kumar, P. Blockchain Technology for Security Issues and Challenges in IoT. *Procedia Comput. Sci.* **2018,** *132*, 1815–1823; https://doi.org/10.1016/j.procs.2018.05.140.
3. Alavi, A. H.; Jiao, P.; Buttlar, W. G.; Lajnef, N. Internet of Things-Enabled Smart Cities: State-of-the-Art and Future Trends. *Measurement* **2018**, *129*, 589–606. https://doi.org/10.1016/j.measurement.2018.07.067.
4. Mehmood, Y.; Ahmad, F.; Yaqoob, I.; Adnane, A.; Imran, M.; Guizani, S. Internet-of-Things-Based Smart Cities : Recent Advances and Challenges. *IEEE Commun. Mag.* **2017,** *55*, 16–24. https://doi.org/10.1109/MCOM.2017.1600514.
5. Pourghebleh, B.; Navimipour, N. J. Data Aggregation Mechanisms in the Internet of Things : A Systematic Review of the Literature and Recommendations for Future Research. *J. Netw. Comput. Appl.* **2017**, *97*, 23–34. https://doi.org/10.1016/j.jnca.2017.08.006.
6. Mourtzis, D.; Vlachou, E.; Milas, N. Industrial Big Data as a Result of IoT Adoption in Manufacturing. *Procedia CIRP* **2016**, *55*, 290–295. https://doi.org/10.1016/j.procir.2016.07.038.
7. Banerjee, M.; Lee, J.; Choo, K. R. A Blockchain Future for Internet of Things Security: A Position Paper. *Digit. Commun. Networks* **2018**, *4* (3), 149–160. https://doi.org/10.1016/j.dcan.2017.10.006.
8. Majeed, M. A. A.; Rupasinghe, T. D. Internet of Things (IoT) Embedded Future Supply Chains for Industry 4. 0: An Assessment from an ERP-Based Fashion Apparel and Footwear Industry. *J. Supply Chain Manag.* **2017,** *6* (1), 24–40.
9. Mineraud, J.; Mazhelis, O.; Su, X.; Tarkoma, S. A Gap Analysis of Internet-of-Things Platforms. *Comput. Commun.* **2016,** *89–90*, 5–16.

10. Bizanis, N.; Kuipers, F. A.; Member, S. SDN and Virtualization Solutions for the Internet of Things : A Survey. *IEEE Access* **2016**, *4*, 5591–5606. https://doi.org/10.1109/ACCESS.2016.2607786.
11. Challa, S.; Wazid, M.; Das, A. K. Secure Signature-Based Authenticated Key Establishment Scheme for Future IoT Applications. *IEEE Access* **2017**, *5*, 3028–3043. https://doi.org/10.1109/ACCESS.2017.2676119.
12. Qiu, T.; Member, S.; Chen, N.; Li, K.; Member, S. How Can Heterogeneous Internet of Things Build Our Future: A Survey. *IEEE Commun. Surv. Tutorials* **2018**, *20* (3), 2011–2027. https://doi.org/10.1109/COMST.2018.2803740.
13. Marjani, M.; Nasaruddin, F.; Gani, A.; Member, S.; Karim, A.; Abaker, I.; Hashem, T.; Siddiqa, A.; Yaqoob, I. Big IoT Data Analytics: Architecture, Opportunities, and Open Research Challenges. *IEEE Access* **2017**, *5*, 5247–5261. https://doi.org/10.1109/ACCESS.2017.2689040.
14. Bibri, S. E. The IoT for Smart Sustainable Cities of the Future: An Analytical Framework for Sensor-Based Big Data Applications for Environmental Sustainability. *Sustain. Cities Soc.* **2018**, *38*, 230–253. https://doi.org/10.1016/j.scs.2017.12.034.
15. Maksimovic, M. *Greening the Future: Green Internet of Things (G-IoT) as a Key Technological Enabler of Sustainable Development Green Economy*; 2018; https://doi.org/10.1007/978-3-319-60435-0.

CHAPTER 2

Design and Analysis of Advanced MOSFET Structures for IoT Applications

SUMAN LATA TRIPATHI[*]

Lovely Professional University, Punjab, India

[*]E-mail: tri.suman78@gmail.com

ABSTRACT

The increasing demand for portable Internet of things (IoT) systems leads designer and engineers to develop new integrated circuit (IC) technology and device structures to meet the requirement for optimum IC area, low consumption, and better speed. Metal oxide semiconductor field effect transistors (MOSFETs) are integral parts of these low-power ICs. So, there is a need to explore new MOSFET architecture with advanced material, structure, and technologies. The design of new MOSFET structures is mainly based on gate or channel engineering with the variations in transistor dimensions, gate oxide or channel materials, and doping of source drain and channel region. The present chapter deals with the fundamental concept to study and analyze new MOSFET structures depending on their low-power applications for IoT-enabled systems.

2.1 INTRODUCTION

The technology enhancement is needed every year depending on various applications and accommodates more transistors per unit area of IC. Since the transistors are basic components of any analog or digital IC, a significant change is required at the level of transistor to meet the requirement of these low-area, low-power integrated circuits (ICs) for Internet of things

(IoT)-enabled systems.[1] The conventional Metal oxide semiconductor field effect transistors (MOSFETs) available in bulk silicon or silicon on insulator (SOI) MOSFET structures are suitable up to 90–180-nm technology node. When there is a decrease in technology below 90 nm, short channel effect (SCE) degrades the device performance and leads to additional heating that can be responsible for nonreliable operation of transistor in the long run. So, there is a need to increase the gate control over the channel to suppress the SCE arising due to high electric field stress between gate and channel/drain region. The gate control can be improved by increasing the number of gates around the channel region. More number of gates results in more gate control over the channel. Multigate MOSFETs are potential candidates that meet the design requirements in low dimension with low power consumptions.[2–4]

2.2 TYPES OF AVAILABLE MOSFET STRUCTURES AND DIMENSIONS

The MOSFETs are available in various shapes and dimensions showing enhanced performance even with low dimension.

2.2.1 BULK MOSFET

MOSFET has evolved with time from single-gate, planar, conventional into double-gate (DG), triple-gate (TG), junctionless, SOI MOSFET, etc. A conventional MOSFET is a device having four terminals: source, drain, gate, and bulk. The gate voltage controls the current flowing from source to drain. The size of the bulk MOSFET is more than 45 nm, and the scaling of this MOSFET is difficult below 32 nm. Further scaling of this MOSFET reduces the distance between source, drain, and channel of the MOSFET producing coupling effect between the channel and source/drain giving rise to SCEs. Figure 2.1 shows the 3D image of bulk MOSFET structure.

2.2.2 SOI MOSFET

SOI technology uses a layer of silicon insulator and a layer of silicon in the substrate instead of a silicon substrate. The parasitic capacitances formed between the source, drain, and the substrate due to depletion of charges are reduced using SOI MOSFET. There are two types of SOI MOSFET: fully depleted SOI MOSET and partially depleted SOI MOSFET. Fully depleted SOI MOSFET has a thin layer of silicon in the upper part of the substrate, and hence, the channel is depleted from the majority charge carriers. The SOI layer is very thin in this structure, hence improving the control of gate over the voltage. The partially depleted SOI MOSFET has thick SOI layer on top and is thicker than the depletion width of the gate (more than 50 nm), which improves the sensitivity and threshold voltage of the device.[6] Figure 2.2 shows 2D view of SOI MOSFET.

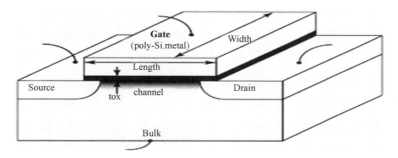

FIGURE 2.1 Structure of bulk MOSFET.

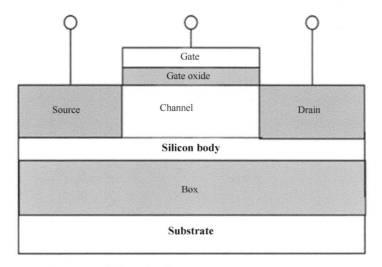

FIGURE 2.2 Structure of SOI MOSFET.

2.2.3 MULTIGATE MOSFET WITH SOI BOX REGION

Multigate MOSFET includes different structures having more than one gate. DG MOSFET has substrate sandwiched between the two gates. Using two gates improve the control of the gate over the depletion region of channel and reduce the SCEs. FinFET is a device that has a silicon body called fin, and the gate is fabricated all around the fin structure. TG MOSFET has gates on three sides around thin silicon film. DG MOSFET provides better scalability, and the subthreshold current reduces, has near ideal subthreshold slope (SS), and less gate leakage current. DG provides high drive current at low power supply; transverse electric field reduces and provides better control over channel.[2] Figure 2.3 shows 2D view of double-gate MOSFET, FinFET, and triple-gate FET respectively.

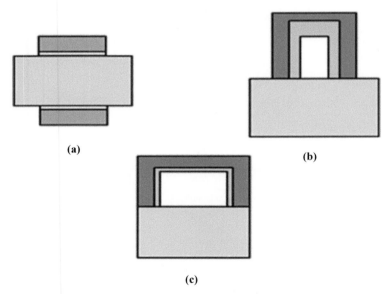

FIGURE 2.3 (a) Double-gate MOSFET, (b) FinFET, and (c) triple-gate FET.

2.2.4 MULTIGATE MOSFET WITHOUT SOI BOX REGION

The multigate MOSFETs with SOI buried oxide region (BOX) have better subthreshold performance and more control of gate on the channel. At the

Design and Analysis of Advanced MOSFET Structures

same time, SOI DG, or TG FinFET suffers from the self-heating effect because of incapability to dissipate heat through SOI insulating layer in comparison to bulk DG or TG MOSFET. Bulk DG or TG MOSFET as shown in Figures 2.4 and 2.5 provide more area to dissipate heat through the large bulk regions. But SOI multigate MOSFET is more suitable in low-power high-speed applications. The speed of bulk transistors can be improved by adding the modifications like bottom spacer (Fig. 2.6 shows a FinFET with bottom spacer) or heavy body doping. So optimization is required in terms of power consumptions with respect to transistor speed.[2]

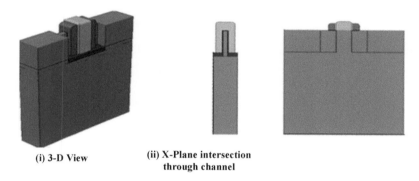

(i) 3-D View   (ii) X-Plane intersection through channel

FIGURE 2.4   3D and cross-sectional view of bulk FinFET.

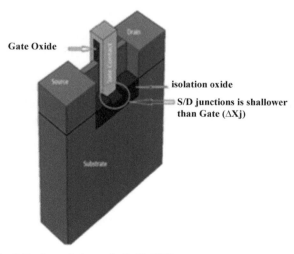

FIGURE 2.5   3-D view of pi-gate bulk FinFET.

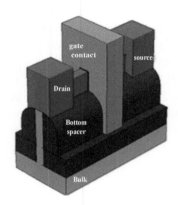 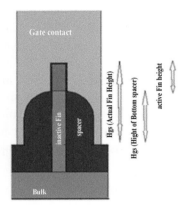

FIGURE 2.6 3-D schematic view of bottom spacer bulk FinFET.

2.2.5 ASYMMETRIC GATE MOSFET

2.2.5.1 ASYMMETRIC DG MOSFET

In asymmetric gate, the top and bottom gates are not aligned. So, there are three types of gates to channel region. The overlap region in the figure is of 10 nm, where the transistor is acting as DGMOSFET. The other two regions are just like normal single-gate MOSFET. Ying Wang proposed a junctionless MOSFET with an asymmetrical gate (AJ MOSFET) to improve the functioning of the device. The AJ MOSFET shown in Figure 2.7 has two gates with a lateral offset between them. In this structure, the channel length depends upon the ON and OFF state of the MOSFET. The channel length of the MOSFET during ON state is equal to overlap length of the gate and the channel length during OFF state is combined length of the two gates minus the overlap length of the gate.

2.2.6 GAA MOSFET[7]

In gate-all-around (GAA) MOSFET, the channel is surrounded by gate from all sides. In GAA, the channel can be cubical or cylindrical shape. Cylindrical GAA MOSFET[7] is also known as nanowire transistors as

Design and Analysis of Advanced MOSFET Structures 25

shown in Figure 2.8. Since there is a circular-shaped gate around the channel, a good control of gate has been projected over the channel region that enhances the overall current drive capabilities of transistor along with better subthreshold performance.

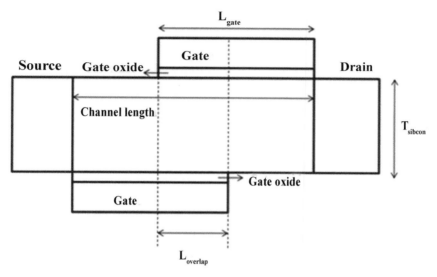

FIGURE 2.7 Asymmetric JLDG MOSFET.

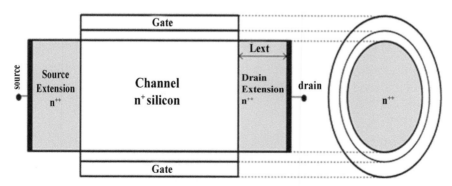

FIGURE 2.8 Cylindrical GAA MOSFET.

2.3 PERFORMANCE ANALYSIS

The performance of MOSFET is mainly decided to drain current versus gate voltage or drain voltage characteristics. The MOSFET performance parameters are mainly categorized as DC and AC parameters.

2.3.1 PERFORMANCE PARAMETERS

The performance parameters are classified as DC and AC parameters. The DC parameters are normally calculated from drain current versus gate voltage characteristics with suitable DC applied voltage at the gate and channel electrodes. The AC parameters are mainly calculated for either single or multiple frequencies by using AC sweep or transient analysis.

2.3.1.1 DC PARAMETERS

2.3.1.1.1 Threshold Voltage

The minimum amount of gate to source voltage required for channel inversion is known as MOSFET threshold voltage (V_T). The value of threshold voltage depends mainly on surface potential, which is the voltage of MOSFET capacitor surface (top layer of polysilicon or metal above the oxide) and voltage in the bulk of MOSFET.

i) OFF-state current

If applied gate voltage is less than channel threshold voltage, the transistor is said to be in OFF-state. So, there is a very small leakage of current flows in OFF-state condition.

ii) ON-state current

It is defined by the value of drain current when the transistor is in ON-state. The ON-state n-channel MOSFET is observed when gate voltage is higher than channel threshold voltage.

Design and Analysis of Advanced MOSFET Structures

iii) Drain-Induced Barrier Lowering (DIBL)

For smaller channel length, the drain electric field stress on the channel will be more, which leads to the variation in threshold voltage. Such reduction in threshold voltage with respect to the change in drain voltage is known as drain-induced barrier lowering (DIBL).

$$DIBL = \frac{V_{Th}^{D} - V_{Th}^{low}}{V_{DD} - V_{D}^{low}} \quad (2.1)$$

iv) SS

It is defined as an inverse of the slope of drain current versus gate voltage. The sharpness of the subthreshold curve indicates the high value of I_{ON}/I_{OFF} current ratio. It also indicates well-defined switching from OFF to ON state.

$$SS = \frac{dV_{gs}}{d(\log I_d)} \; (mV/decade) \quad (2.2)$$

$$Or, \; SS_{th} = \ln(10)\frac{kT}{q}\left(1 + \frac{C_d}{C_{ox}}\right) \quad (2.3)$$

2.3.1.2 AC PARAMETERS

2.3.1.2.1 Analog/Radio Frequency (RF) Performance Parameters

The analog and RF performance mainly depends on transconductance, transistor capacitances, and frequency response.

i) Transconductance

Transconductance (gm) is defined as the ratio of change in drain current with respect to the change in gate voltage of transistor.

$$g_m = \frac{2I_{DS}}{|V_P|}\left(1 - \frac{V_{GS}}{V_P}\right) \quad (2.4)$$

ii) Capacitances

The MOSFET acts like a capacitor. The total capacitance of MOSFET can be defined approximately equal to the sum of gate to source and gate to drain capacitances.

$$C_T = C_{gs} + C_{gd} \tag{2.5}$$

2.4 FET-BASED MEMORY DESIGN

Advanced MOSFET structures can be implemented in memory design to achieve target of low leakage and low power consumption to support future scaling trends with an increased number of component per chip for large storage capacity SRAM and DRAM in IoT-enabled systems. SRAM utilizes four transistors to store 1-bit information, the structure of which has two cross-couple inverters. Further two transistors are utilized as control access to the put-away information during read and write operation. The SRAM cell has three distinct states. In view of various conditions of SRAM cell control, scattering in FinFET circuits can be arranged into two primary parts:

1) Static power dissipation: when the circuit is in idle mode

Dynamic power dissipation: When performing read–write operations. It depends on the number of transitions during these operations.

The scaling of transistor and voltages introduces additional burden on IC in terms of leakage current resulting in more static power dissipation. So, use of advanced MOSFET like FinFET overcomes this issue with low-leakage power dissipation and enhanced subthreshold performance. Figure 2.9 presents a 7T SRAM architecture using FinFET. A comparison has been made that 7T FinFET SRAM cell shows less power dissipation than that of 7T CMOS SRAM cell as shown in Table 2.1.

TABLE 2.1 Power Comparison of SRAM Cell.

Operation	7T CMOS SRAM (45-nm CMOS) (nW)	7T FinFET SRAM (18-nm FinFET) (nW)
Write0	25.32	20.2
Write1	25.86	21.89
Read0	25.54	20.80
Read1	25.22	19.58

Design and Analysis of Advanced MOSFET Structures

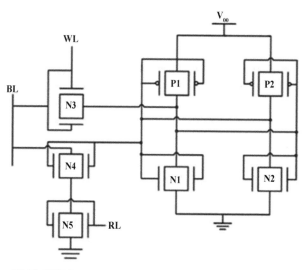

FIGURE 2.9 7T FinFET SRAM cell.

A 6T SRAM cell with junctionless FET as shown in Figure 2.10 was designed with less IC area and low power consumption per unit IC. The area taken by junctionless FET-based 6T SRAM cell was presented is 6.9 μm² in comparison to conventional structure is 11.3 μm² with power consumption under specified limit. Hence, we can add more number of SRAM cells with such low-power transistors on same silicon chip, that is, the requirement of large-capacity memory architectures.

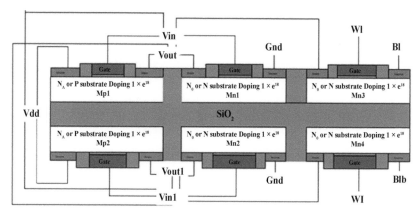

FIGURE 2.10 Architecture of 6T SRAM cell with junctionless MOSFET.

2.5 BIOMEDICAL APPLICATIONS

FET-based biosensors work on the electrical property of the device (i.e., ON-state current, threshold voltage and capacitances, etc.). In dielectrically modulated FET (DM FET), a cavity is formed in the gate dielectric medium to restrict the movement of charged and noncharged biomolecules for the detection of molecules that are label free. The biomolecules trapped inside the cavity region change the electrical parameters of the FET.[8–10] Biosensors based on DM FET have some limitations like SCEs, problems related to scaling and power supply. Early-stage detection of disease is easily possible using FET-based biosensors. A nanocavity was introduced in the gate oxide region of the MOSFET using the process of etching from both sides of the device near the drain and source.[11] The molecules trapped in the cavity region bind to the SiO_2, and surface potential in the cavity region gets affected as shown in Figure 2.11.

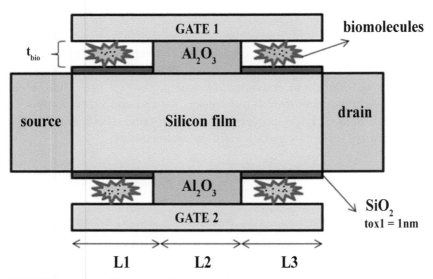

FIGURE 2.11 Double-gate MOSFET with biosensing cavity region.

2.6 CONCLUSION

The new MOSFET structures explored are based on suitable applied gate and channel engineering, including the concept of multigate like DG,

TG, and GAA structures. As the number of gate increases, the control of gate over the channel increases with increasing transistor drive current. Multiple-gate concepts are also suitable for steep subthreshold curve for sub 45-nm technology nodes. Several new MOSFET structures are analyzed for DC and AC parameters based on digital, memory, and analog/RF applications. The junctionless transistors are found suitable in sub-20-nm technology nodes with low voltage and high I_{ON}/I_{OFF} current ratio. DG MOSFET structure can also be developed as biosensors by the addition of cavity regions between gate and channel interface. These cavity regions are sensitive to biomolecule variations that reflect in terms of dielectric properties of gate and channel interface.

KEYWORDS

- **low power**
- **memory design**
- **biomedical**
- **junctionless**
- **SCE**
- **IoT**

REFERENCES

1. Tripathi, S. L. Low Power High Performance Tunnel FET: Analysis for IOT Applications. In *Handbook of Research on the Internet of Things Applications in Robotics and Automation*; IGI Global Publisher, 2019; DOI: 10.4018/978-1-5225-9574-8.ch002; ISBN:9781522595748.
2. Colinge, J. P. Multi-gate SOI MOSFETs. *Microelectron. Eng.* **2007,** *84*, 2071–2076.
3. Tripathi, S. L.; Mishra, R.; Mishra, R. A. In *Characteristic Comparison of Connected DG FINFET, TG FINFET and Independent Gate FINFET on 32 nm Technology*, IEEE ICPCES, 2012, pp 1–7.
4. Tripathi, S. L.; Mishra, R.; Narendra, V.; Mishra, R. A. In *High Performance Bulk FinFET with Bottom Spacer*, IEEE CONECCT, 2013; pp 1–5.
5. Patel, G. S.; Tripathi, S. L.; Awasthi, S. In *Performance Enhanced Unsymmetrical FinFET and Its Applications*, IEEE EDKCON, 2018, pp 222–227.
6. Mendiratta, N.; Tripathi, S. L. A Review on Performance Comparison of Advanced MOSFET Structures Below 45 nm Technology Node. *J. Semiconduc., IOP Sci. (Scopus)* **2020,** *41*, 1–10.

7. Choudhary, P.; Kapoor, T. Structural and Electrical Analysis of Various MOSFET Designs. *J. Eng. Res. Appl.* **2015,** *5* (3), 16–19.
8. Djeffal, F.; Ferhati, H.; Bentrcia, T. Improved Analog and RF Performances of Gate-All-Around Junctionless MOSFET with Drain and Source Extensions. *Superlattices Microstruct.* **2016,** *90,* 193.
9. Kumari, V.; Saxena, M.; Gupta, R. S.; Gupta, M. Two Dimensional Analytical Drain Current Model for Double Gate MOSFET Incorporating Dielectric Pocket. *IEEE Trans. Electron. Devices* **2012,** *59,* 2567.
10. Narang, R.; Saxena, M.; Gupta, M. Comparative Analysis of Dielectric-Modulated FET and TFET-Based Biosensor. *IEEE Trans. Nanotechnol.* **2015,** *14* (3), 427–435.
11. Tripathi, S. L.; Patel, R.; Agrawal, V. K. Low Leakage Pocket Junction-Less DGTFET with Bio Sensing Cavity Region. *Turkish J. Electric. Eng. Comput. Sci.* **2019,** *27* (4), 2466–2474.
12. Ajay, N. R.; Saxena, M.; Gupta, M. Investigation of Dielectric Modulated (DM) Double Gate (DG) Junctionless MOSFETs for Application as a Biosensors. *Superlattices Microstruct.* **2015,** *85,* 557.
13. Mishra, V. K.; Chauhan, R. K. Efficient Layout Design of Junctionless Transistor Based 6-T SRAM Cell Using SOI Technology. *ECS J. Solid State Sci. Technol.* **2018,** *9,* 456.

CHAPTER 3

Integration of MEMS Sensors for Advanced IoT Applications

ANUJ KUMAR GOEL*

Department of Electronics and Communication Engineering, Chandigarh University, Punjab, India

*E-mail: anuj40b@gmail.com

ABSTRACT

Connecting many devices and systems via the Internet makes them smart by the use of bidirectional information sharing and by the shared control of system. Many sensing devices are connected in Internet of Things (IoT) to make them smarter and capable to take independent decisions. Advanced sensors are designed and fabricated using microengineering, which is the most recent advancement in electronics industries. It ranges from few millimeters to few nanometers with applications in every aspect of social and economic life. Main industrial devices developed using MEMS that are used in IoT systems are soil moisture sensor, temperature sensor, gyroscopes, accelerometers, biosensors, pressure sensors, magnetometers, optical actuator, gas sensors, etc. These miniature devices are very helpful due to their compact size and minimum power requirement as compared to macrodevices. MEMS (micro electro mechanical systems) devices find their place in IoT industry with their low power and less space requirements.

3.1 FUNDAMENTALS OF MEMS TECHNOLOGY

MEMS is an advanced fabrication methodology, a novel method for fabricating electromechanical devices with the use of batch fabrication technique like the ICs are fabricated and manufacturing these electromechanical essentials all along through electronics technology.[1,2] This technology enabled novel approaches in science and engineering, for example, polymerize chain reaction microsystems regarding DNA amplification and recognition, the micromachined scanning tunneling microscopes, biochips regarding detection of hazardous and choice of the engineering segment.[3–6]

The characteristics of MEMS devices and their diversifications in various domains make them potentially a more persistent technology than even VLSI (very large scale integration) microchips. Nowadays, cost and reliability of MEMS devices are comparable with VLSI devices, because this technology permits complicated electromechanical systems to design and fabricate by means of batch fabrication. With regular advancements in MEMS industry, the price of MEMS devices is predicted to be much lower as compared to macroscale components and systems.[7,8]

As advanced technique permitting unparalleled collaboration among hitherto unrelated branches of endeavor such as biotechnology and microengineering, this technique is predicted to comprise a business and defense marketplace enlargement as closely related to integrated circuit technologies. Many foreign countries such as United States, Japan and many Europe governments invested huge amount of money for the research, innovation, and marketable application of MEMS appliances.[9–11]

This technology will be without doubt coming footprint in electronics industry with integration of electronics and mechanical systems and creates the aspiration of fabricating systems smaller, for example, modern vehicles. Due to the last few decades of intense scientific research and innovation of integrated circuits, nowadays, manufacturing industries have approximately every essential apparatus and methodologies desirable for complete fabrication of MEMS systems.[12] That is why, creating advancements in microengineered devices has become moderately simpler and inexpensive.

MEMS devices are capable in implementation of many sensing mechanisms, for example, physical and chemical sensing, actuation, and communication.[13–16] Main focus in MEMS is on its two main properties: (1) micro/nano size and (2) assurance of low production cost.[17,18]

Different mathematical equations are explained later to demonstrate the behavior of the device. First is relation between beam's natural frequency, spring constant, and mass of device. Natural frequency is that value on which cantilever freely oscillates without external force. While

Hook's law defines the spring constant that is a characteristic of material, in microcantilevers, main dominating parameter is natural frequency which depends on k and m as represented by eq 3.1.[19-22] Hence, with the increase in mass (m), the natural frequency is reduced. For better natural frequencies, material with higher spring constant (k) is chosen.

$$\omega_0 = \sqrt{\frac{k}{m}} \tag{3.1}$$

The relation for spring constant is mentioned in eq 3.2. It is directly proportional to the thickness and the width of beam and inversely proportional to the length of beam. Hence, with increase in thickness and width, the parameter gets higher value, and at the same time, enhanced length reduces the same. These relations are well proved by simulations of various beams with the changes in respective physical dimensions.

$$k = \frac{Et^3 w}{4l^3} \tag{3.2}$$

Relation between beam deflection (z) and differential surface stress (δs) is mentioned in eq 3.3. With increase in the length of beam, the stress on fixed end is increased and with increase in the thickness of beam, the deflection is reduced.

$$z = \frac{3L^2(1-\vartheta)}{3t^2}\delta s \tag{3.3}$$

3.2 DEFINING THE INTERNET OF THINGS

In Internet of Things (IoT), all connected devices are divided in three segments:

(a) devices that gather information and forward them,
(b) devices that take delivery of information and do something on it, and
(c) devices that perform both duties.

Abovementioned categories include colossal advantages that multifaceted one on the top of other.

(a) Gathering and Forwarding Information

Different MEMS sensors are integrated in IoT to intellect and measure physical quantities, that is, humidity, temperature, moisture, air quality, and light. All sensors, in a system, automatically gather information from the outer environment which make possible to take better and accurate decisions.[23–25]

In agriculture, automatically gathering information by the soil moisture sensor makes farmers capable to decide when to start watering for better crop production.[26] By advanced IoT algorithms, the farmers can decide not to water too much (which is a wastage of expensive natural resource) or less (i.e., crops production reduced). By using these advanced sensing mechanisms, farmers can increase their crop yield as well as there can be decrease in usage of natural resources.

Not only machines but humans also are benefited, as in human beings the five senses make individual capable for having knowledge about environment, similarly, MEMS devices make IoT enough intelligent with knowledge about environment.

(b) Delivering and Do Something on Information

In this category, machines get the information and then act on it. For example, water pump receives the electrical signal and watering starts. Temperature sensor receives a high alert signal and then starts controlling the environment temperature.[27–30] IoT systems make peripheral devices capable to take every type of decision either simple one as to sense humidity in air or complex as human behavior monitoring.

Full efficiency of IoT is achieved, when devices perform combined duties as (a) and (b), that is, gathering and forwarding information and delivering and doing something on that information.

(c) Performing Both: The Goal of an IoT System

In the farm field example, the sensors can accumulate data for water content in the farm and humidity in the environment to alert farmers regarding watering the crop, although the farmers are practically not needed. MEMS-integrated IoT devices can work by their own based on how much water content present in the farm and humidity present in the environment.

Also, rain sensor gives signal to irrigation system about weather conditions. So based on that feedback mechanism, irrigation system can decide, not to water the crops because it will be watered by the rain anyways.

In addition to previous features, all the data regarding soil moisture, how much watering of farm is done by the irrigation system, and crop monitoring is controlled by IoT and carry forward to servers, where optimized logics take intelligent decisions with all gathered data.

3.2.1 MAIN ELEMENTS OF IOT SYSTEM

3.2.1.1 SENSORS AND ACTUATORS

First of all, sensing devices gather data from surroundings. The received information from sensors can be very straightforward that humidity measurement otherwise may be complicated as human behavior monitoring.

A number of sensing elements can be used in a system to measure many parameters and also can take cumulative decisions to get optimum result. To illustrate as, mobile handsets are integrated with various MEMS sensors to get audio, video signals, acceleration, and location detection, still mobile handsets are never considered as only sensing element because they are doing sensing of many elements.[31–33]

However, whether it is a standalone sensor or a full device, in this first step, data is being collected from the environment by something.

3.2.1.2 CONNECTIVITY

After the first stage, connectivity is utilized for data transmission to cloud. In connectivity, proper methodology of algorithms is required.

Various technologies that include cellular, satellite, Wi-Fi, Bluetooth, linking through a gateway or router, or direct connection to the Internet are used in IoT system. All these technologies are integrated with sensing devices that are further coupled with cloud.[34,35]

All technologies have ups and downs among power consumption, range, and bandwidth. Depending on the specific IoT application, any of the parameter can be superimposed over other, but all are performing the similar job: transmitting information to the cloud.

3.2.1.3 DATA PROCESSING

Optimized algorithms are used for data processing of received signals in cloud proceeding stage 1. This process can be very straightforward, as to check the humidity either in predefined limit or not. Otherwise the processing may also be complicated (thefts on a property).

3.2.1.4 USER INTERFACE

Last but not the least, gathered data should be available to the user on the other end by various methodologies. It is possible by the means of email, text, notification, etc. as an alert to the user. For example, GSM (global system for mobile communication) enabled SMS (short message service) alert when the soil is too dry in the farm field.[36]

End users have crossing points that allow users to efficiently monitor connected surroundings. To illustrate, farmers can check soil moisture level, humidity, pH level, etc. by the means of mobile handset applications or Internet surfing.

This user interface can be unidirectional. Depends on a particular IoT application, the user on the monitoring side is capable for performing a task and to change system variables. For example, the farmer can power on the water pump in the farm field from the distant location by the means of an application on the mobile phone or by giving just a miss call on predefined GSM number.

Some activities can be automatically adjusted, instead of coming up of user to correct temperature; the IoT arrangement with the help of preprogrammed criteria can perform this automatically. Instead of calling the user to make attentive regarding an intruder, this IoT methodology can automatically inform safety members or appropriate establishment.[37,38]

3.3 MEMS MICROCANTILEVERS

These are basic elements, designed like rectangular beams with dimensions varying from micrometer to nanometer range. Presently, researchers are fabricating these in various shapes and sizes that depend on application.

Different materials are used to fabricate these devices with much advancement. Starting from polysilicon, aluminum, and silicon nitride followed by SU8 are used in present sensing devices.

Figure 3.1 shows the microbeam with predefined dimensions. The physical dimensions can be varied, and respective change in strain can be measured. Due to their micro size and high sensitivity, these are used in physical, chemical, and biosensing applications.

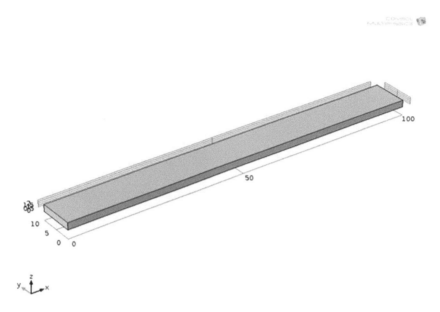

FIGURE 3.1 Design of microbeam.

While designing the devices, first, we have to set the physics of the device. Second, material properties are applied to device. Third, loading mechanism is formulated according to sensing mechanism. It is followed by meshing of the device to distribute the device in many elements. FEA (finite element analysis) is the basic principle of MEMS simulation that divides the geometry in many elements to obtain optimized result. COMSOL Multiphysics, Ansys, IntelliSuite, CoventorWare, etc. are leading tools used for designing and simulation of MEMS and NEMS devices.

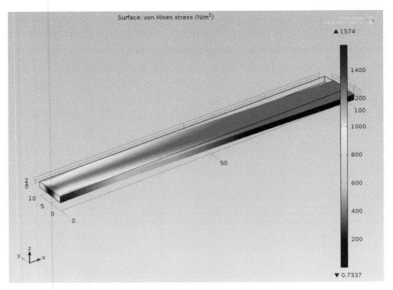

FIGURE 3.2 Simulated device design with result.

After meshing of the device, final simulation is performed to get result with defined parameters.

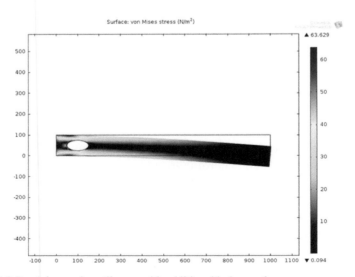

FIGURE 3.3 Advanced cantilevers with additional hole creation.

Figures 3.2 and 3.3 are simulated models without any hole formation and with hole formation at fixed end. With change in structure of device, respective change in resultant model is obtained, which can further be enhanced to get better results. Figure 3.4 demonstrates the same model with different types of hole formation. As stress concentration regions are changing, respective change in resultant stress and strain is measured.

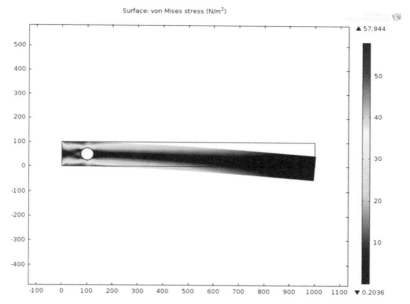

FIGURE 3.4   Revised Figure 3.3 model with different hole formation.

Further microcantilevers are fabricated with many hole formation on fixed end as shown in Figure 3.5. With reduced mass of device, the resultant stress is increased. In fabrication industry, different approaches are used to increase efficiency of device in terms of detection of minute particles (Fig. 3.5).

3.3.1   ELECTROSTATIC ACTUATOR

One application of the abovementioned device is electrostatic actuator and model is shown in Figure 3.6. Potential difference of 5 V is applied between two electrodes. As a result, electrostatic force is generated and beam deflection is measured. With implementation of advanced microbeams, effective deflection is increased, which means that they are more sensitive to same amount of applied potential (Fig. 3.6).

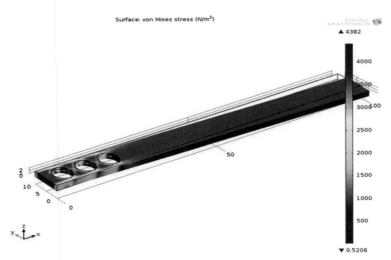

FIGURE 3.5 Final model view with more number of holes on fixed end.

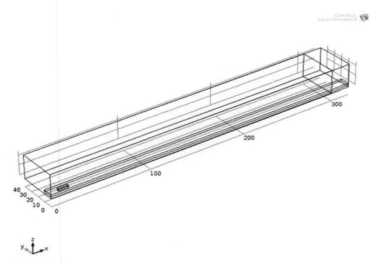

FIGURE 3.6 Electrostatic actuator with microbeams.

Integration of MEMS Sensors for Advanced IoT Applications

As application of advanced microbeams gives more deflection and it can be measured by respective increase in capacitance as shown in Figure 3.7. Change in capacitance occurs due to the change in the position of beams.

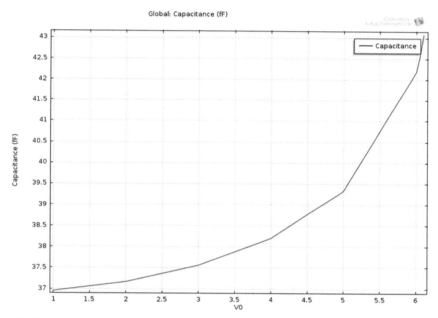

FIGURE 3.7 Capacitance plot with potential change.

3.4 GAS SENSOR WITH MICRO DIMENSIONS

Gas sensing is done by change in surface acoustic waves (SAW). By the transduction principle, the applied signal that is electrical in nature is converted to wave because mechanical behavior is more affected by physical processes. This sensing element afterward transduces the mechanical quantity reverse in initial signal that is an electrical signal. Gas sensing is done by monitoring of abrupt variations in electrical characteristics of signal that is amplitude, phase, frequency, or time-delay between the input and output electrical signals.

Travelling waves in SAW device are affected by the change of mass over surface of sensing element as waves are travelled across the delay line. Relation between "Young's modulus E" and "density of material" is given by equation:

$$v = \sqrt{\frac{E}{\rho}} \tag{3.4}$$

For that reason, factor "v" reduces with increase in device mass. The velocity variation comes due to SAW mechanism that is predicted by respective time-delay or phase shift between input and output signals. With decrease in wave's energy, the signal attenuation can also be measured due to combination with the extra surface mass. For the case of mass sensing, as the change in the signal will always be due to an increase in mass from a reference signal of zero additional mass, signal attenuation can be effectively used.

The sensor consists of an interdigitated transducer etched onto a piezoelectric substrate, covered with a thin film. When material is adsorbed on the surface of sensor then the mass of the film increases as its material selectively adsorbs a chemical substance from the air. These mass changes cause a deviation in resonance to a slightly lower frequency and proves the detection of impurities in the sample. Figure 3.8 show modeled SAW device for gas-sensing application. This model possesses one microcantilever in its geometry. Figure 3.9 demonstrates same device with two microcantilevers.

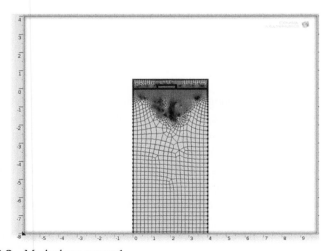

FIGURE 3.8 Meshed gas sensor 1.

Integration of MEMS Sensors for Advanced IoT Applications 45

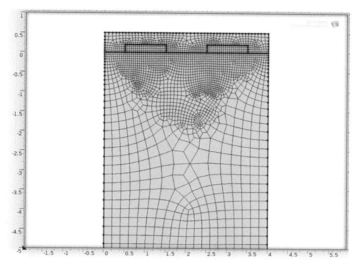

FIGURE 3.9 Meshed gas sensor 2.

Simulated models are demonstrated for designed two SAW device models as Figures 3.10 and 3.11.

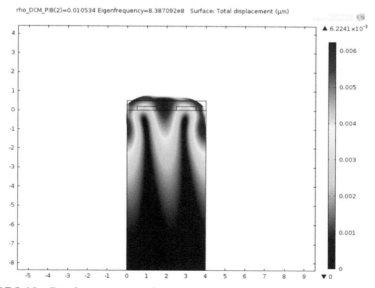

FIGURE 3.10 Resultant gas sensor 2.

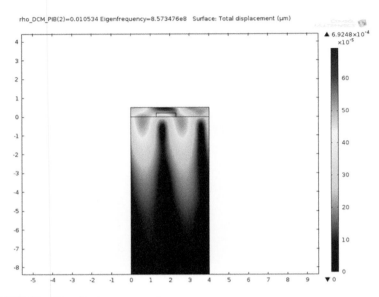

FIGURE 3.11 Resultant gas sensor 1.

Figures 3.12 and 3.13 represent the variation of Eigen frequency and change in potential difference on SAW designed models.

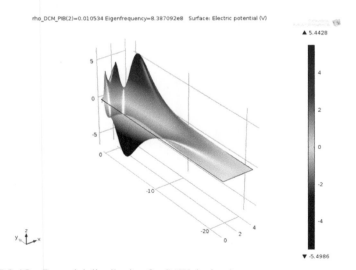

FIGURE 3.12 Potential distribution for SAW device 1.

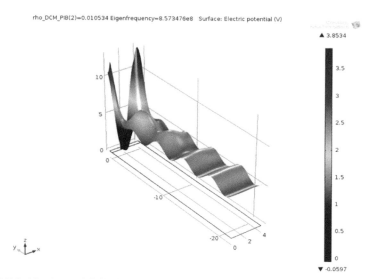

FIGURE 3.13 Potential distribution for SAW device 2.

The MEMS sensors are used in IoT systems to collect signals from their outer environment and process the signals in an effective way to communicate to further connecting devices. For example, multigas sensor finds its applications in industries where there is chance of gas leakage, in health care to sense harmful gases, in smart cities to provide optimum environmental conditions, in smart homes to sense harmful gases, required oxygen level, in farming to sense harmful gases for crop enhancing, etc. Figure 3.14 illustrates various application areas of gas sensors.

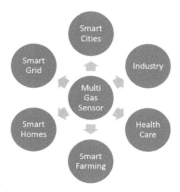

FIGURE 3.14 IoT potential markets.

3.5 CONCLUSION

Commercially many advanced MEMS sensing devices are available that are used in IoT applications. "AS-MLV-P2 MEMS MOS" Gas sensor, "GM-402B-24 V CH4" MEMS Combustible gas sensor, and "Alcoa Prime MiCS5524CO" Alcohol and VOC gas sensors are the few sensors that are available in the market. Sensor industry is filled with microengineered sensors that take over the macroscale technologies. IoT systems become smart and intelligent with use of MEMS-based sensing elements. Machine learning is another highlight in electronics industry that will defiantly enhance applications of IoT appliances.

KEYWORDS

- **MEMS devices**
- **IoT**
- **microcantilevers**
- **MEMS gas sensor**

REFERENCES

1. Clark, J. Self-Calibration and Performance Control of MEMS with Applications for IoT. *Sensors* **2018**, *18*, 1–19.
2. Ciuti, G.; Ricotti, L.; Menciassi, A.; Dario, P. MEMS Sensor Technologies for Human Centred Applications in Healthcare, Physical Activities, Safety and Environmental Sensing: A Review on Research Activities in Italy. *Sensors* **2015**, *15*, 6441–6468.
3. Sparks, D. MEMS Packaging for the IoT. *EPEM* **2017**, *262*, 35–38.
4. Vashist, S. K. A Review of Microcantilevers for Sensing Applications. *J. Nanotechnol.* **2007**, *3*, 1–15.
5. Taj, M.; Vikram, N. V. R. G. MEMS-Based Energy Harvesters for IoT Applications. *IJRTE* **2019**, *7*, 433–436.
6. Goel, A. K.; Reddy, B. S. K. Performance Analysis of SAW Gas Sensors with Different Number of Electrodes. **2019**, *IJRTE 8*, 4397–4401.
7. Grayson, A. C. R.; Shawgo, R. S.; Johnson, A. M.; Flynn, N. T.; Li, Y.; Cima, M. J.; Langer, R. A BioMEMS Review: MEMS Technology for Physiologically Integrated Devices. *Proc. IEEE* **2004**, *92*, 6–21.
8. Chaterjee, S.; Pohit, G. A Large Deflection Model for Pull-In Analysis of Electrostatically Actuated Microcantilever Beam. *J. Sound Vibr.* **2009**, *322*, 964–986.

9. Goel, A. K. Modern Electronics Wearable Gadgets for Health Monitoring. *RTSRT* **2019**, *6*, 11–16.
10. Chen, G. Y.; Thundat, T.; Wachter, E. A.; Warmack, R. J. Adsoption Induced Surface Stress and Its Effects on Resonance Frequency of Microcantilevers. *J. Appl. Phys.* **1995**, *77*, 3618–3622.
11. Chowdhury, S.; Ahmadi, M.; Miller, W. C. A Closed-Form Model for the Pull-In Voltage of Electrostatically Actuated Cantilever Beams. *J. Micromech. Microeng.* **2005**, *15*, 756–763.
12. Goericke, F. T.; King, W. P. Modeling Piezoresistive Microcantilever Sensor Response to Surface Stress for Biochemical Sensors. *IEEE Sens.* **2008**, *8*, 1404–1408.
13. Goel, A. K. Analytical Modeling and Simulation of Microcantilever Based MEMS Devices. *Wulfenia J.* **2017**, *24*, 79–91.
14. Goel, A. K.; Kumar, K.; Gupta, D. Design and Simulation of Microcantilevers for Sensing Applications. *Int. J. Appl. Eng. Res.* **2016**, *11*, 807–809.
15. Greminger, M. A.; Sezen, A. S.; Elson, B. J. A Four Degree of Freedom MEMS Microgripper with Novel Bi-Directional Thermal Actuators. *Proc. Int. Conf. Intell. Robots Syst.* **2005**, 2814–2819.
16. Haluzan, D. T.; Klymyshyn, D. M.; Achenbach, S.; Börner, M. Reducing Pull-In Voltage by Adjusting Gap Shape in Electrostatically Actuated Cantilever and Fixed-Fixed Beams. *Micromechanics* **2010**, *1*, 68–81.
17. Joglekar, M. M.; Pawaskar, D. N. Closed—Form Empirical Relations to Predict the Dynamic Pull-In Parameters of Electrostatically Actuated Tapered Microcantilevers. *J. Micromech. Microeng.* **2011**, *21*, 1–12.
18. Lo, C. C.; Meng-Ju, L.; Chang-Li, H. Modeling and Analysis of Electrothermal Actuators. *J. Chin. Inst. Eng.* **2009**, *32*, 351–360.
19. Rosminazuin, A. R.; Badriah, B.; Burhanuddin, Y. M. Design and Analysis of MEMS Piezoresistive SiO_2 Cantilever-Based Sensor with Stress Concentration Region for Biosensing Applications. *ICSE* **2008**, 211–215.
20. Tao, F.; Wang, Y.; Zuo, Y.; Yang, H.; Zhanga, M. Internet of Things in Product Life-Cycle Energy Management. *J. Ind. Inf. Integr.* **2016**, *1*, 26–39.
21. Tragos, E. Z.; Foti, M.; Surligas, M.; Lambropoulos, G. An IoT Based Intelligent Building Management System for Ambient Assisted Living. *IEEE Int. Conf. Commun. Workshop* **2015**, 8–12.
22. Saptasagare, V. S. Next of Wi-Fi a Future Technology in Wireless Networking Li-Fi Using Led Over Internet of Things. *Int. J. Emerg. Res. Manage. Technol.* **2014**, *3*.
23. Granjal, J.; Monteiro, E.; Sá Silva, S. Security for the Internet of Things: A Survey of Existing Protocols and Open Research Issues. *IEEE Commun. Surv. Tutorials* **2015**, *17*, 1294.
24. Atzori, L.; Iera, A.; Morabito, G. The Internet of Things: A Survey. *Comput. Netw.* **2010**, *54*, 2787–2805.
25. Perera, C.; Zaslavsky, A. B.; Christen, P.; Georgakopoulos, D. Context Aware Computing for the Internet of Things: A Survey. *IEEE Commun. Surv. Tutorials* **2014**, *16*, 415–454.
26. Perera, C.; Zaslavsky, A. B.; Christen, P.; Georgakopoulos, D. Sensing as a Service Model for Smart Cities Supported by Internet of Things. *Trans. Emerg. Telecommun. Technol.* **2014**, *25*, 81–93.

27. Zanella, A.; Bui, N.; Castellani, A.; Vangelista, L.; Zorzi, M. Internet of Things for Smart Cities. *IEEE IoT J.* **2014**, *1*, 22–32.
28. Perera, C.; Zaslavsky, A.; Christen, P.; Georgakopoulos, D. Context Aware Computing for the Internet of Things: A Survey. *IEEE Commun. Surv. Tutorials* **2014**, *16*, 414–454.
29. Gil, D.; Ferrandez, A.; Mora-Mora, H.; Peral, J. Internet of Things: A Review of Surveys Based on Context Aware Intelligent Services. *Sensors* **2016**, 1–23.
30. Donohoe, M.; Jennings, B.; Balasubramaniam, S. Context-Awareness and the Smart Grid: Requirements and Challenges. *Comput. Netw.* **2015**, *79*, 263–282.
31. Sen, S. Context-Aware Energy-Efficient Communication for IoT Sensor Nodes. *IEEE Des. Autom. Conf.* **2016**, 1–6.
32. Sanchez, L.; Munoz, L.; Galache, J. A.; Sotres, P.; Santana, J. R.; Gutierrez, V.; Ramdhany, R.; Gluhak, A.; Krco, S.; Theodoridis, E.; Pfisterer, D. Smart Santander: IoT Experimentation Over a Smart City Testbed. *Comput. Netw.* **2014**, *61*, 217–238.
33. Magno, M.; Polonelli, T.; Benini, L.; Popovici, E. A Low Cost, Highly Scalable Wireless Sensor Network Solution to Achieve Smart LED Light Control for Green Buildings. *IEEE Sens. J.* **2015**, *15*, 2963–2973.
34. Evans, D. The Internet of Things: How the Next Evolution of the Internet Is Changing Everything. Whitepaper, April 2011. https://www.cisco.com/c/dam/en_us/about/ac79/docs/innov/IoT_IBSG_0411FINAL.pdf.
35. Xu, K.; Qu, Y.; Yang, K. A Tutorial on the Internet of Things: From a Heterogeneous Network Integration Perspective. *IEEE Netw.* **2016**, *30*.
36. Zorian, Y. In *Ensuring Robustness in Today's IoT Era*, Design & Test Symposium (IDT), 2015 10th International, 2015, pp 14–16.
37. Saffre, F. In *The Green Internet of Things*, Innovations in Information Technology (IIT), 2015 11th International Conference on, 1–3 Nov. 2015.
38. Ray, S.; Jin, Y.; Raychowdhury, A. The Changing Computing Paradigm with Internet of Things: A Tutorial Introduction. *IEEE Des Test* **2016**, *33*.

CHAPTER 4

CMOS Bootstrap Driver

ABHISHEK KUMAR*

*School of Electronics and Electrical Engineering,
Lovely Professional University, Punjab, India*

*E-mail: abkvjti@gmail.com

ABSTRACT

Bootstrapped driver circuit operates at ultra-low supply voltage. Bootstrapping is a technique to improve the driving capability of the digital circuit to operate on very low voltage. The conventional bootstrap-driven circuit operates at 1–1.5 V; in the present work the supply voltage reduces up to 0.8 V at CMOS 180 nm and 0.3 V at CMOS 90 nm. The final output of the driver boosted the above supply voltage. The presented driver circuit shows a low leakage current dissipates minimum power, resulting in 81.46% improvement in power dissipation and 91.2% improvement in delay.

4.1 INTRODUCTION

Complementary metal oxide semiconductor (CMOS)-based logic gate is the default standard for the electronic design automation (EDA) tool. The power consumption of a circuit heavily relies on the technology used in the design. The trend to reduce technology lowers the power requirement. If the CMOS technology scales down to 180, 90, and 45 nm, supply voltages lower to 1.8, 1, and 0.7 V, respectively (Karthikeyan and Mallick, 2019). The threshold voltage of MOS devices is a critical parameter of the design; it provides a range of energy to consider logic 0 and logic 1. Reduction of supply reduces the threshold voltage too. Each cell library has defined the lower limit of the threshold voltage. Body biasing is an alternative to play with a threshold voltage. Usually, the fourth terminal of MOS devices is grounded; biasing at the fourth terminal can increase or decrease the threshold voltage. A driver's circuit design is the main component to determine the overall system performance. VLSI circuit consists of a large capacitive load (Kuo, 2004). With an increase and decrease in gate voltage, NMOS and PMOS devices are bootstrapping indirectly, mainly used to derive sizeable capacitive load. It is a challenging task to acquire the high capacitive load with low supply voltage (Sundararajan and Parhi, 1999; Lou, 1997; Rofail, 1999). The chapter deals with the indirect bootstrap technique of gate deriver for a large capacitive load at the output when the bootstrap technique at gate output of the device is not impressive in reducing the transition speed of the output. The structure of the chapter is organized as follows. Section 4.2 puts light on the bootstrap technique. Section 4.3 presents the modified bootstrap driver result that is analyzed in Section 4.4.

4.2 CONCEPT OF BOOTSTRAPPING

Bootstrapping is referred to as pulling up the physical variables. Here, we will explain the situation in which voltage on separate (isolated) node is increased to the high power supply value level V_{DD} due to dynamic switching.

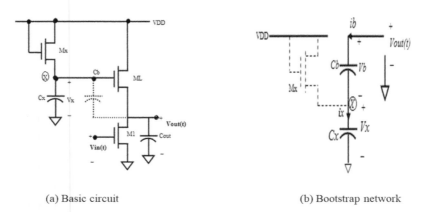

(a) Basic circuit (b) Bootstrap network

FIGURE 4.1 (a) Basic bootstrapping circuit and (b) bootstrap networks.

CMOS Bootstrap Driver

The bootstrap mechanism can be analyzed as shown in Figure 4.1(a). Here, in this circuit, input $V_{in}(t)$ can be considered as square input that switches between 0 V and V_{DD}. Here, switching transistor M1 acts as a logic device (Yeo et al., 2000; Garcia, 2006; Kuo, 2006); $V_{out}(t)$ is in the range of V_{OL}–V_{OH} as a set of DC transfer properties of the circuit. The presented circuit is different from another course as V_{OH} is a function of time because of the bootstrapping applied at the voltage at internal node V_x. To demonstrate activity of this circuit, we will extract DC characteristics. If V_{in} is 0 V, M1 will be in cutoff, and MOS "MX" acts like a coupled pass transistor (PT) among supply rail V_{DD}, and terminal "X" will give the value of voltage V_x across C_x as shown in eq 4.1.

$$V_X = V_{DD} - V_{TX} \qquad (4.1)$$

where V_{TX} is the threshold of transistor "M_X," "Cx" is parasitic capacitance onto node "X" measured output.

$$V_{out} = V_X - V_{TL} \qquad (4.2)$$

where V_{TL} is a threshold voltage level of transistor "ML." By ignoring the body biasing effect, eq 4.2 is modified to eq 4.3.

$$V_{OH} = V_{DD} - V_{TX} - V_{TL}$$
$$= V_{DD} - 2V_T \qquad (4.3)$$

where V_T is a threshold voltage. $V_{in} = V_{DD}$ calculates the low output voltage. It will turn ON the transistor M1 pull-down output voltage to GND, while if $V_{out} = V_{OH}$, the M1 transistor is not saturated. While the voltage V_x is measured as $(V_{DD} - V_{TX})$, if the transistor "ML" gets saturated, the output current will be as eq 4.4

$$\frac{\beta_{n1}}{2}[2(V_{DD} - V_{T1})V_{OL} - V_{OL}^2] = \frac{\beta_n L}{2}(V_x - V_{TL})^2 \qquad (4.4)$$

Eq 4.4 is a quadratic equation, which can solve for V_{OL} depending upon aspect ratio, and V_{OL} will always be higher compared to 0 (gnd).

Bootstrapping technique is the active event that occurs due to mutual combination between the output voltage V_{out} and node X with the help of bootstrap capacitor C_b. Figure 4.1(b) shows the bootstrap network, which includes C_b and C_x. C_x is the node capacitance of node X, and C_b is the value of bootstrap capacitance in the circuit addition of parasitic

G–S terminal capacitance contribution of transistor "ML." When dynamic switching is observed, limitation by measurement shows that V_{out} will rise and fall periodically with the change in input voltage $V_{in}(t)$. The significant features can be included by first observing the current flow in the circuit as shown in Figure 4.1(b), and current flow through C_x is

$$ix = Cx\frac{dVx}{dt} \tag{4.5}$$

When the current flowing into capacitor C_p is

$$ib = Cb\frac{d(Vout - Vx)}{dt} \tag{4.6}$$

As $i_b = i_x$, the result obtained is

$$Cx\frac{dVx}{dt} = Cb\frac{d(Vout - Vx)}{dt} \tag{4.7}$$

$$\frac{dVx}{dt} = \left(\frac{Cb}{Cb + Cx}\right)\frac{dVout}{dt} \tag{4.8}$$

So, eq 4.8 shows that voltage at internal node $V_x(t)$ is coupled dynamically to $V_{out}(t)$. We can study bootstrap dynamics by assuming initial condition as $V_{in} = V_{DD}$, so that $V_{out}(0) = V_{OL}$. When the voltage input changes to $V_{in} = 0$ V, output voltage rises with time because transistor "ML" charges C_{out}. This indicates that $(dV_{out}/dt) > 0$, so that (dV_x/dt) is also greater than 0, which means $V_x(t)$ enhances. By multiplying both sides by "dt," integrate both sides (Kim and Kong, 2008; Garcia, 2004; Kong and Jun, 1999; Wong et al., 2010).

$$\int_{(VDD-VTX)}^{Vx(t)} dVx = \left(\frac{Cb}{Cb + Cx}\right)\int_0^t \left(\frac{dVout}{dt}\right)dt \tag{4.9}$$

$$Vx(t) = (VDD - VTX) + r\int_0^t \left(\frac{dVout}{dt}\right)dt \tag{4.10}$$

where we have included coupling ratio "r" as a parameter that highlights the energy level of coupling. The best combination occurs when $C_b \gg C_x$ and $r \approx 1$.

CMOS Bootstrap Driver

$$r = \left(\frac{Cb}{Cb + Cx}\right) \quad (4.11)$$

$$Vx(t) = (VDD - VTX) + r\left[Vout(t) - VOL\right] \quad (4.12)$$

We will integrate eq 4.10 and $V_{out}(0) = V_{OL}$; this will allow us to calculate $V_x(t)$. As transistor ML is PT during charging event, $V_{out}(t)$ is measured using device eq 4.13 to solve transient response. The output of the circuit in Figure 4.1(a) shows that

$$Vgsl = Vx - Vout \quad (4.13)$$

$$Vdsl = VDD - Vout \quad (4.14)$$

since

$$Vx = VDD - VTX \quad (4.15)$$

$$Vsat = Vds - VTX - VTL \quad (4.16)$$

This highlights that $V_{dsl} > V_{sat}$ is true, so transistor "ML" charges C_{out} in conduction in the saturation region. The current will give

$$\frac{\beta nl}{2}(VDD - VTX - VTL - Vout)^2 = Cout\frac{dVout}{dt} \quad (4.17)$$

So, from this equation V_{out} can be calculated using integral

$$\int_{VOL}^{Vout(t)} \frac{dVout}{(VDD - VTX - VTL - Vout)^2} = \int_0^t \frac{\beta nL}{2Cout} dt \quad (4.18)$$

Interchanging and rearranging eq 4.18 will give

$$Vout(t) \approx Vm\left[\frac{\frac{t}{2\tau L}}{1 + \frac{t}{2\tau L}}\right] + \frac{VoL}{1 + t/2\tau L} \quad (4.19)$$

where V_m is the highest output voltage presented by eq 4.20 and the time constant presented by eq 4.21

$$Vm(t) = Vx(t) - VTL = VDD - VTX - VTL \quad (4.20)$$

and

$$\tau L = \frac{Cout}{\beta nL(VDD - VTX - VTL - VOL)} \qquad (4.21)$$

4.2.1 TRADITIONAL BOOTSTRAP TECHNIQUE

Figure 4.2 presents the traditional bootstrapping of a CMOS model, a value of sizeable capacitive load (Luo, 1997). It is observed that the bootstrap circuit consists of pull-up and pull-down sections. Capacitor C_p and NMOS transistor N_{1b}/N_{2b} bootstrap to pull up. Capacitor C_n PMOS transistor P_{1b}/P_{2b} bootstraps the pull-down section.

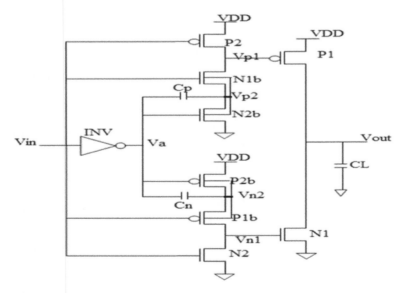

FIGURE 4.2 Traditional bootstrap driver.

The bootstrap technique operation is used to derive a heavy load C_L. While circuit is pulled up, the transient before ramp-up period, C_p will charge to V_{DD}. Stored charge in C_p, upon built-up time period P_1, will be driven by negative gate voltage. Here, a fast pulling up switching is (transition) obtained; it is known as the bootstrap technique. The bootstrap technique method is used to derive a large capacitive load driver circuit. While the circuit is pulled down, switching (transient) circuit explains the similar method to pull up the operation. Bootstrap technique is an alternative method to design low-voltage CMOS logic.

4.2.2 BOOTSTRAP TECHNIQUE IN DYNAMIC CMOS LOGIC

Figure 4.3 shows a 2-input bootstrap CMOS AND logic gate using bootstrapped dynamic logic (BDL) circuit (Garcia, 2004, 2006). Static CMOS BDL-AND gate circuit consists of the CMOS dynamic logic and the bootstrap circuit.

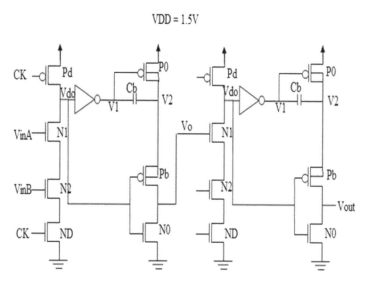

FIGURE 4.3 Bootstrap dynamic CMOS AND logic (BDL-AND).

The BDL-AND gate contains two periods of precharge and evaluation, while in the precharge operation, clock period is low, capacitor C_b is charging upward up to V_{DD}. While in evaluation phase, when the clock is high, with input V_{DD}, due to the charge stored into the capacitor C_b, the voltage at internal node V_2 is bootstrapped to a height above supply voltage V_{DD}, capable of driving the output strength.

4.2.3 BOOTSTRAP TECHNIQUE IN STATIC CMOS LOGIC

As mentioned earlier, the bootstrap technique is used in dynamic logic circuits; it can be used to design static logic circuits by enhancing overall system performance. Here, it is described how the bootstrap technique is used in static circuits with the application of the input gate in the output transistors of the driver. This indirect way of bootstrapping shown in Figure 4.4 is not an efficient technique in the reduction of the transition (switching) speed (Chen and Kuo, 2003; Hush and Thomann, 1992; Garcia, 2007).

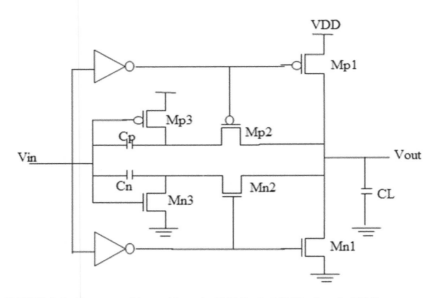

FIGURE 4.4 Bootstrap driver with static CMOS sub-1 V (Gogl et al., 2005).

In direct bootstrapping capacitor, C_p and C_n are connected at output node via transistors MP2 and MN2, instead of gate terminal of the output terminal in contrast to bootstrap indirect method. Direct bootstrap method stores charge in capacitor C_p and C_n before pull up and pull down, the voltage of output terminal can be pulled upward and downward quickly (Nakamizo et al., 2015).

4.3 BOOTSTRAP DRIVER

4.3.1 TRADITIONAL BOOTSTRAP

Figure 4.5 shows CMOS circuit of the traditional bootstrap driver, which is designed with a fundamental segment and bootstrap segment (Kong and Jun, 1999; Gogl et al., 2005; Nakamizo et al., 2015). The fundamental section requires 2 PMOS (PM0 and PM1) and 2 NMOS (NM0 and NM1). The bootstrap section requires 2 PMOS (PM2 and PM3) and 2 NMOS (NM2 and NM3). Transistors of NOT gate "PM4" and "NM4" and bootstrap load "C_0" and "C_1" are included in the bootstrap.

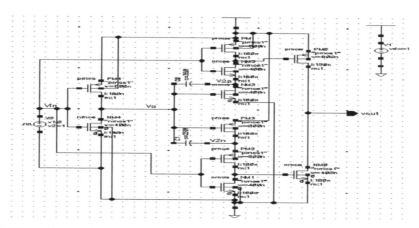

FIGURE 4.5 Traditional bootstrap driver circuit.

Transistors NM_2 and NM_3, along with bootstrap load C_0 during transient, are pulled upward. Transistors PM_2 and PM_3 and capacitor load C_1 are pulled downward. While MOS are pulled up, the traditional bootstrap driver is divided into two periods (Garcia, 2009). First, C_0 is charged in the build-up period and second, in the bootstrap period. When input V_{in} is low, then the output of inverter is high, which will turn NM2 and NM1 OFF and NM3 ON. The level output of traditional bootstrap driver V_{out} driven indirectly by PM_2 results in the bootstrap segment to 0 V. Contrastingly, NM_3 bootstrap load C_0 of the bootstrap section is isolated from PM_1 and PM_0 of the basic section. Bootstrap load C_0 has boosted to 1 C, the left half pulled upward to 1 V, and the right half is node "V_{2p}" to 0 V; NM3 is turned ON (Karthikeyan and Mallick, 2019; Dougherty et al., 2016; Jung et al., 2017).

For input voltage, "V_{in}" = 1 V, inverter output changes down to 0.001 V; the right half of bootstrap load C_0 loses connection from ground terminal since MOS NM_3 is turned OFF. The right half of the bootstrap load C_0 is networked to the gate terminal of PM_0, due to the reason that NM_2 is turned ON. Due to variation in voltage level that occurs in the left half of bootstrap load C_0, there is a decrease of voltage level from 1 to 0.001 V, then the right side of load C_0 will reduce from 0 to −0.506 V. The voltage on V_2 positive terminal reduction exhibits undershoots. The peak of undershoots is estimated by the ratio of bootstrap capacitor C_0 to the parasitic capacitance at the right side of C_0. Undershoots of the output voltage transit at a fast rate since the gate voltage of transistor PM_0 is driven at −0.506 V. The output voltage rises to the full curve of 1.00001 V. Pull-down impermanent is the inverse nature of pull-up impermanent (Sharma, 2017). A traditional bootstrap circuit of bootstrap capacitors is C_0 = 350 fF and C_1 = 250 fF and PMOS width is kept at 800 nm and length at 180 nm, whereas NMOS width is kept 400 nm and length at technology node 180 nm.

4.3.2 LIMITATION OF TRADITIONAL BOOTSTRAP DRIVER

The limitation of traditional bootstrap driver is that it will not activate well if the supply voltage is lower than 1 V. Further, with reduction in supply voltage, the real functionality of driver circuit is modified. Traditional bootstrap requires a transistor count, which helps the circuit areas to get enhanced. The traditional bootstrap driver requires large values of bootstrap capacitor, which further increases area and power requirement. To solve the problem of a traditional bootstrap driver circuit, a modified bootstrap driver is presented in this work. Modified bootstrap driver lowers the capacitor size, supply voltage, and the number of transistors. A traditional bootstrap driver circuit requires a large capacitor size C_p = 350 fF and C_n = 250 fF, while in a modified bootstrap driver, value of bootstrap load capacitance reduces C_p = C_n = 17 fF. The supply voltage of the traditional bootstrap driver is 1 V, and the output is boosted to 1.0001 V, whereas modified bootstrap driver boosts output to 1.4 V. In this work, modified driver is designed using Cadence front end tool at CMOS 180- and 90-nm technology and their performance is analyzed. Bootstrap driver

CMOS Bootstrap Driver

circuit at 180-nm technology with a supply voltage of 0.8 V increases output at the positive level up to 1.17 V. In comparison, CMOS 90-nm technology supply voltage of 0.3 V boosts the positive level to 0.599 V, which is almost double of V_{DD}.

4.3.3 MODIFIED BOOTSTRAP DRIVER

Modified bootstrap driver is designed by using CMOS 180-nm technology with Cadence Virtuoso schematic composer. The minimum input supply voltage for a modified bootstrap driver is 800 mV, as shown in Figure 4.6. In the modified bootstrap driver circuit, the period of the input signal is 20 ns and pulse width is 10 ns; it requires two bootstrap capacitors $C_p = C_n = 17$ fF, which results in output voltage that boosts to 1.2 V. The modified bootstrap capacitors are C_p and C_n, and transistors are "PM1" and "NM2." "C_p" is precharged and "C_n" is predischarged. V_{2p} and V_{2n} are the boosted nodes. The output from transistors "PM3" and "NM3" acts as inverter output, and the output V_a control is from transistors "PM2" and "NM1." The final output voltage of the modified bootstrap driver is boosted above V_{dd} and below GND. When V_{in} is changing from high to low, initially node V_{2n} is at 0 V. When $V_{in} = 0$ V, then node V_{2n} will be boosted to -587.718 mV; here transistor "PM2" is turned OFF and "NM1" turned ON. The boosted signal at node V_{2n} passes from "NM1" and reaches V_{out}. "NM2" is turned ON causing the reverse current flow to charge V_{2n}, and when V_{in} is low, at the end of period node, V_{2n} holds the charge of -250.193 mV.

When input voltage V_{in} is rising, PM_1 turns OFF voltage as node V_{2p} increases above V_{DD} up to 1.4 V. On last point of clock, it holds up to 1.197 V. When input voltage V_{in} is 0.8 V, PM_2 is ON and NM_1 is OFF. Voltage at node V_{2p} increases to 1.475 V and rises to V_{out} with the help of MOSFET PM_2, and NM_2 gets ON. Voltage at terminal V_{2p} precharges capacitor C_n to -12.9994 µV, which is lower than the ground. Output voltage is on complete period of V_{in} when it switches from low to high or high to low. If V_{in} is 0.8 V, V_{out} is boosted above a supply voltage of 1.3 V. If V_{in} is 0 V, V_{out} is boosted below ground terminal -366.843 V. Transistor sizing is done as PMOS width = 800 nm, length = 180 nm and NMOS width = 400 nm, length = 180 nm. The layout of the modified bootstrap driver is designed with Cadence Assura with bootstrap capacitors of 17 fF. These two bootstrap capacitors are implemented with MIMCAP with a width 4 µm and length 4 µm, which achieves minimum area. For increasing the capacitor value, area is increased proportionally. The modified bootstrap driver layout area = 13.345 × 12.745 µm = 170.082025 × 10^{-12} m².

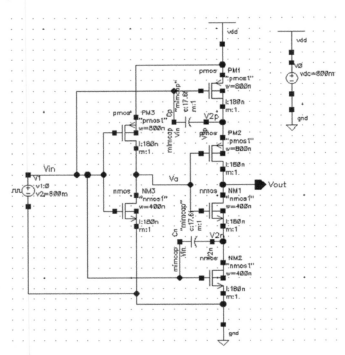

FIGURE 4.6 Modified bootstrap driver.

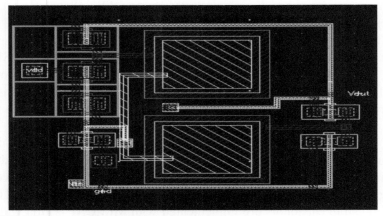

FIGURE 4.7 Layout of modified bootstrap driver.

CMOS Bootstrap Driver

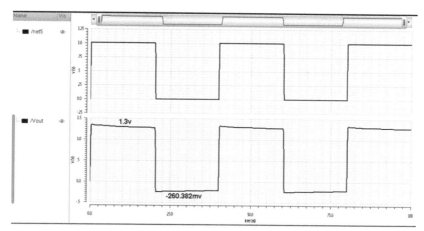

FIGURE 4.8 Transient response of modified bootstrap driver.

4.4 RESULT AND DISCUSSION

Figures 4.7 and 4.8 present the layout and transient response of the modified bootstrap-driven circuit. During high pulse, output level is boosted to 1.2 V and lower level is boosted down to −0.26 V. The effect of technology scaling over output level is depicted in Figure 4.9. For similar supply voltage, bootstrap driver at 90-nm technology is higher compared to other driver circuit as discussed in this chapter. From Figure 4.8, it is observed that the modified bootstrap driver provides better results when compared to the traditional bootstrap driver with smaller technology nodes.

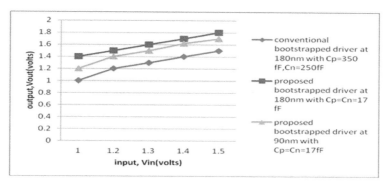

FIGURE 4.9 Output level of bootstrap driver circuits.

Figure 4.10 shows the comparison of power dissipation in various bootstrap drivers. A traditional bootstrap at CMOS 180 nm dissipates a power of 9.73 µW. A modified bootstrap driver at 180 nm dissipates power value estimated as 65.7 µW, and for the modified bootstrap driver at 90 nm, power consumption value is 68.3 µW.

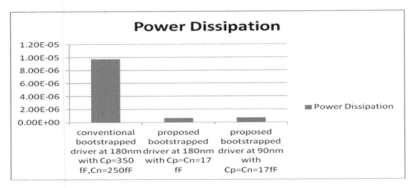

FIGURE 4.10 Comparison of power dissipation in various bootstrap driver circuits.

Figure 4.11 shows the comparison of different delay bootstrap drivers. A traditional bootstrap driver at CMOS 180 nm possesses a delay of 20.08 ns. A modified bootstrap driver at 180 nm possesses a delay of 20.07 ns, and in a modified bootstrap driver at CMOS 90 nm, the delay is of 20.0 ns.

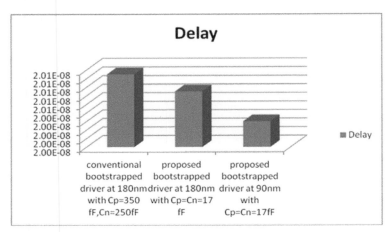

FIGURE 4.11 Comparison of delay in different bootstrap driver circuits.

CMOS Bootstrap Driver

Figure 4.12 presents the comparison of the current leaks that occur in different bootstrap drivers. A traditional bootstrap driver possesses current leakage to 3.50 pA, while in a modified bootstrap driver at CMOS 180 nm, current leakage value is 3.49 pA, and in a modified bootstrap driver circuit at CMOS 90 nm, current leakage is 97.8 pA.

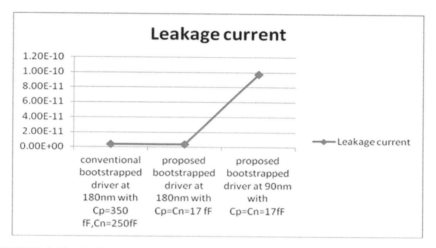

FIGURE 4.12 Leakage current of various bootstrap.

Leakage current that increases with technology scaling, for similar capacitive load leakage current, is higher for bootstrap driver circuit at CMOS 90-nm technology compared to CMOS 180 nm. The modified bootstrap driver is faster compared to the traditional ones. Similarly, the power dissipation of the modified bootstrap driver is lower compared to the traditional one as shown in Table 4.1.

TABLE 4.1 Bootstrap Driver Parameter.

Driver	CMOS technology (nm)	Capacitor	PMOS width (nm)	NMOS W/L (nm)	PMOS counts	NMOS counts
Traditional bootstrap	180	$C_p = 350$ fF, $C_n = 250$ fF	800	400	5	5
Modified bootstrap driver	180	$C_p = C_n = 17$ fF	800	400	3	3
Modified bootstrap driver	90	$C_p = C_n = 17$ fF	360	120	3	3

In this chapter, we explored different ways to design low supply voltage CMOS bootstrap driver circuit. Here, we have explained the bootstrapping technique and analyzed its usefulness. It can be used in different circuits. The modified driver circuit provides better results when compared to the traditional driver in terms of power, delay, and leakage current. Modified bootstrap CMOS driver schematic is implemented with Cadence Virtuoso schematic composer at CMOS 180- and 90-nm technologies; it requires lower area as highlighted in the layout. Bootstrap driver's application is found in charge pump and logic circuit. A modified bootstrap charge pump circuit gives boosted output as compared to the traditional bootstrap charge pump. Modified bootstrap logic gates also offer increased production as compared to traditional logic gates.

4.5 CONCLUSION

In this chapter, bootstrapped techniques are shown and explained how to use them in different circuits. The proposed driver circuit gives better results compared to conventional driver circuit with low power dissipation, less delay, and less leakage current. Proposed bootstrapped driver circuits are designed at Cadence Virtuoso 180-nm as well as 90-nm technologies. From layout of proposed bootstrapped driver circuit, it is verified that proposed driver circuit occupies less area. The proposed bootstrapped charge pump circuit gives boosted output as compared to conventional bootstrapped charge pump. Proposed bootstrapped logic gates also give boosted output as compared to conventional logic gates.

KEYWORDS

- **BDL**
- **driver**
- **bootstrap**
- **reduced supply**
- **leakage current**

REFERENCES

Chen, J. H. T.; Kuo, J. B. Ultra-low-voltage SOI CMOS Inverting Driver Circuit Using Effective Charge Pump Based on Bootstrap Technique. *Electron. Lett.* **2003**, *39* (2), 183–185.

Chen, Y.; Geng, D.; Lin, T.; Mativenga, M.; Jang, J. Full-Swing Clock Generating Circuits on Plastic Using a-IGZO Dual-Gate TFTs with Pseudo-CMOS and Bootstrapping. *IEEE Electron Device Lett.* **2016**, *37* (7), 882–885.

Dougherty, C. M.; Xue, L.; Pulskamp, J.; Bedair, S.; Polcawich, R.; Morgan, B.; Bashirullah, R. A 10 V Fully-Integrated Switched-Mode Step-up Piezo Drive Stage in $0.13\,\upmu\text{m}$ CMOS Using Nested-Bootstrapped Switch Cells. *IEEE J. Solid-State Circ.* **2016**, *51* (6), 1475–1486.

García, J. C.; Montiel-Nelson, J. A.; Sosa, J.; Navarro, H. A Direct Bootstrap CMOS Large Capacitive-load Driver Circuit. *Proc. Design Auto. Test Eur. Conf. Exhib.* **2004**, *1*, 680–681.

García, J. C.; Montiel-Nelson, J. A.; Nooshabadi, S. A Single-capacitor Bootstrap Power-efficient CMOS Driver. *IEEE Trans. Circ. Syst. II: Exp. Briefs* **2006a**, *53* (9), 877–881.

García, J. C.; Montiel-Nelson, J. A.; Nooshabadi, S. Bootstrap Full-swing CMOS Driver for Low Supply Voltage Operation. *Proc. Design Auto. Test Eur. Conf.* **2006b**, *1*, 2–5.

García-Montesdeoca, J. C.; Montiel-Nelson, J. A.; Nooshabadi, S. DB-driver: A Low Power CMOS Bootstrap Differential Cross-coupled Driver. *Int. J. Electron.* **2007**, *94* (9), 809–819.

García, J. C.; Montiel-Nelson, J. A.; Nooshabadi, S. In *Analysis and Comparison of High-Performance CMOS Adiabatic Drivers*, IEEE International Symposium on Circuits and Systems; 2009a; pp 3146–3149.

García, J. C.; Montiel-Nelson, J. A.; Nooshabadi, S. In *High Performance CMOS Dual Supply Level Shifter for a 0.5 V Input and 1 V Output in Standard 1.2 V 65 nm Technology Process*, 9th International Symposium on Communications and Information Technology; 2009b; pp 963–966.

Gogl, D.; Arndt, C.; Barwin, J. C.; Bette, A.; DeBrosse, J.; Gow, E.; Maffitt, T. A 16-Mb MRAM Featuring Bootstrap Write Drivers. *IEEE J. Solid-State Circ.* **2005**, *40* (4), 902–908.

Hush, G.; Thomann, M. R. U.S. Patent No. 5,128,563; U.S. Patent and Trademark Office: Washington, DC, 1992.

Jung, G.; Tekes, C.; Pirouz, A.; Degertekin, F. L.; Ghovanloo, M. In *Beyond Supply-Voltage Bootstrapped Pulser for Driving CMUT Arrays in Ultrasound Imaging*. IEEE Biomedical Circuits and Systems Conference; **2017**; pp 1–4.

Karthikeyan, A.; Mallick, P. S. Body-Biased Subthreshold Bootstrapped CMOS Driver. *J. Circ. Syst. Comput.* **2019**, *28* (03), 1950051.

Kim, J. W.; Kong, B. S. Low-Voltage Bootstrap CMOS Drivers with Efficient Conditional Bootstrapping. *IEEE Trans. Circ. Syst. II: Exp. Briefs* **2008**, *55* (6), 556–560.

Kong, B. S.; Jun, Y. H. Power-efficient Low-voltage Bootstrap CMOS Latched Driver. *Electron. Lett.* **1999**, *35* (24), 2113–2115.

Kong, B. S.; Kang, D. O.; Jun, Y. H. In *A Bootstrap CMOS Circuit Technique for Low-voltage Application*, ICVC'99. 6th International Conference on VLSI and CAD, (Cat. No. 99EX361); 1999; pp 289–292.

Kuo, J. B.; Chen, P. C. Sub-1 V CMOS Large Capacitive-load Driver Circuit Using Direct Bootstrap Technique for Low-Voltage CMOS VLSI. *Electron. Lett.* **2002,** *38* (6), 265–266.

Kuo, J. B.; Lin, S. C. *Low-Voltage SOI CMOS VLSI Devices and Circuits*; John Wiley & Sons, 2004; pp 269–279.

Lou, J. H.; Kuo, J. B. In *1.5 V CMOS and BiCMOS Bootstrap Dynamic Logic Circuits Suitable for Low-Voltage VLSI*, Proceedings of Technical Papers. International Symposium on VLSI Technology, Systems, and Applications; **1997a**; pp 279–282.

Lou, J. H.; Kuo, J. B. A 1.5-V Full-Swing Bootstrap CMOS Large Capacitive-Load Driver Circuit Suitable for Low-Voltage CMOS VLSI. *IEEE J. Solid-State Circ.* **1997b,** *32* (1), 119–121.

Nakamizo, H.; Mukai, K.; Shinjo, S.; Gheidi, H.; Asbeck, P. In *Over 65% PAE GaN voltage-mode Class D Power Amplifier for 465 MHz Operation Using Bootstrap Drive* IEEE Topical Conference on Power Amplifiers for Wireless and Radio Applications; **2015**; pp 1–3.

Rofail, S. S. Sub-1V Bootstrap CMOS Driver for Giga-scale-integration Era. *Electron. Lett.* **1999,** *35* (5), 392–394.

Sharma, V. K. Design of Low Leakage PVT Variations Aware CMOS Bootstrapped Driver Circuit. *J. Circ. Syst. Comput.* **2017a,** *26* (09), 1750137.

Sharma, V. K. In *Low Leakage Circuit Design Using Bootstrap Technique*; Fourth International Conference on Signal Processing, Communication and Networking; **2017b**; pp 1–4.

Sundararajan, V.; Parhi, K. K. In *Low Power Synthesis of Dual Threshold Voltage CMOS VLSI Circuits*, Proceedings of the International Symposium on Low Power Electronics and Design; 1999; pp 139–144.

Wong, O. Y.; Tam, W. S.; Kok, C. W.; Wong, H. In *A Low-voltage Charge Pump with Wide Current Driving Capability*, IEEE International Conference of Electron Devices and Solid-State Circuits; 2010, pp 1–4.

Yeo, K. S., Ma, J. G.; Do, M. A. Ultra-Low-Voltage Bootstrap CMOS Driver for High Performance Applications. *Electron. Lett.* **2000,** *36* (8), 706–708.

CHAPTER 5

Implementation of Low-Power BIST Using Bit Swapping Complete Feedback Shift Register (BSCFSR)

RAVI TRIVEDI[1] and SANDEEP DHARIWAL[2*]

[1]*Digicomm Semiconductors Pvt. Ltd., Bengaluru 560103, Karnataka, India*

[2]*Department of ECE, Alliance College of Engineering and Design, Alliance University, Bengaluru 562106, Karnataka, India*

Corresponding author. E-mail: dhariwal.vlsi@gmail.com

ABSTRACT

Design for Testability (DFT) has been a major concern for today's VLSI engineers. A poorly designed DFT would result in losses for manufacturers with a considerable rework for the designers. BIST (Built-in Self-Test)—nne of the promising DFT techniques is rapidly modifying with the advances in technology as devices shrink. Because of the growing complexities of the hardware, the trend has shifted to include BISTs in high performance circuitry for offline as well as online testing. Work done here involves testing a Circuit Under Test (CUT) with built-in response analyzer and vector generator with a monitor to control all the activities. Use of low transition vector generators here Bit-Swapping Complete Feedback Shift Register (BS-CFSR), power has been reduced considerably when compared to classical Linear Feedback Shift Register (LFSR) techniques. This presents the process of design implementation for a complete BIST working on both normal operation mode as well as test mode for any 4 bits' circuitry. Xilinx ISE 14.7 for coding in Verilog and implementation with Cadence's Encounter[(R)] RTL Compiler RC 14.10 was used for timing responses, with power consumption calculated at different technology nodes. In IoT applications, reliability of electronic devices plays a crucial role. A faulty design cannot support reliable components and sensors. Therefore BS-CFSR will be best BIST structure to test and rectify any fault occurrence at system level.

5.1 INTRODUCTION

Internet of things (IoT) is used in a wide range of applications, including portable and handy devices, medicine, transportation, and smart homes. Logic BIST (Built-in Self-Test) is also crucial for many of these similar applications, for example, defense, automotive, banking, computer, health-care, networking, and telecommunication industries. Reliability of the system can be enhanced by the best testing measures. The availability of low-power BIST using bit swapping complete feedback shift register (BS-CFSR) for electronic devices is one of the main foundations for IoT applications. With opportunities, there are several challenges in the design of VLSI IoT devices. Meanwhile, the devices become more portable and handy; power reduction in today's VLSI circuits is a key parameter step for designing and testing. Interestingly, the power consumed in testing process is higher than that in normal operation. Four major reasons assumed for power rise when the circuit being tested are[1-3]

- rapid transition of the test bits,
- simultaneous activation of the whole circuit during test,
- extra test circuitry, and
- low correlation among test vectors.

A major concern for today's VLSI engineers is testability. An inefficient DFT leads to redesign for the designers and eventually it is a significant loss to the industry. BIST is helping in debugging the designs rapidly, with the latest technology compatibility as device shrinks. Because of the growing complexities of the hardware, the trend has shifted to test the device for offline and online BIST enable testing.[4-6] In this research work, a circuit under test (CUT) with built-in response analyzer is undertaken. This CUT design is enabled with vector generator with activities controller. BS-CFSR is one of the best transition vector generators. This BS-CFSR is used to reduce the power considerably when compared to classical "linear feedback shift register" (LFSR) techniques.[7–9] This presents the process of design implementation for a complete BIST working on both normal operation mode as well as test mode for any 4-bit circuitry. Xilinx ISE 14.7 for coding in Verilog and implementation with Cadence's Encounter® RTL Compiler ripple carry (RC) 14.10 was used for timing responses, with power consumption calculated at different technology nodes. IoT-based devices depend a lot on the real-time performance. A faulty design cannot support reliable components and sensors. Hence, BS-CFSR is the best self-test design to debug and diagnose maximum faults.

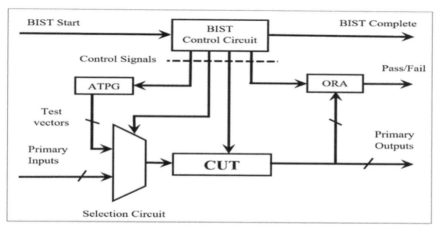

FIGURE 5.1 Basic BIST blocks.[8]

A DFT methodology can be applied at the specific circuit for checking faults and produces a response saying if it has passed the test or not using BIST structure.[2] This self-test feature comes with a trade-off—more area requirement and higher power consumption are due to the extra circuit added to the functional circuit. The CUT is considered in combination with ATPG (Automatic Test Pattern Generation) and ORA (Output Response Analyzer). One of the important TPGs, that is, LFSR, is used to generate patterns acting as pseudorandom test patterns. Modifications are performed on the classical model to improve any of the abovementioned problems. One of them is used in proposed BIST implementation. The output response compression techniques and signature analysis are some of ORAs that lessen the output response and makes it easier for the controller to give results.[2] The so-called golden vectors (fault-free responses) are stored in ROM memory of the circuit. Figure 5.1 shows basic blocks building a BIST.

5.2 METHODOLOGY

5.2.1 BS-CFSR

LFSR is series of flops connected in a fashion to generate test patterns in every clock cycle. Ex-OR gates are used for feedback path, and depending on where these gates are connected, there can be internal LFSR or external LFSR. The tap sequences (Ex-OR gate sites) chosen help to produce almost all possible vectors, called primitive polynomial. Some resistant fault sites need all zero vectors to sensitize, where in classical model cannot be used. A modified LFSR is shown in Figure 5.2. This is incorporated with addition of gates and flip-flops to get 4-bit CFSR.[5]

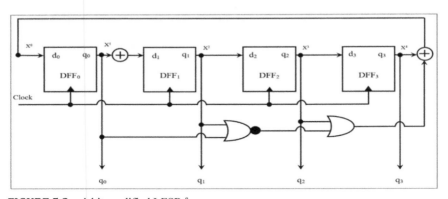

FIGURE 5.2 4-bit modified LFSR.[8]

The pseudorandom nature of CFSR increases the fault coverage in a relatively short span of test vectors, but this process also increases the transition counts and number of switches in the CUT. This in turn raises the power consumption during test mode, which can lead to problems of power capacity and reliability.[7] This directs to modify the CFSR structure that leads to rectify itself to minimize the transition in the vectors generated without compromising the fault coverage. The revised CFSR is one that reduces the number of transitions in the CUT test vector inputs by 25%

using a bit-swapping technique without compensating on randomness of the test patterns. The area overhead is extra 2:1 multiplexers for at least $n-2$ flops. If an n-bit CFSR with any seed value executes for 2^n clock cycles and give test vectors, then transitions equal to 2^n are produced at the final output of CFSR. Modification is done by considering one of CFSR's outputs (assume nth bit) to be a control line. Conditions are if n is odd bit and bit $n =$ zero (0); in this way, first bit will be swapped with second bit, third with bit number four, …, bit $n-2$ with bit $n-1$. If n is even and bit $n = 0$, then bit 1 will be swapped with bit 2, bit 3 with bit 4, …, bit $n-3$ with bit $n-2$. If bit $n=1$, no swapping is initialized. In all cases of swapping operation, the control line "bit n" is not included.[3]

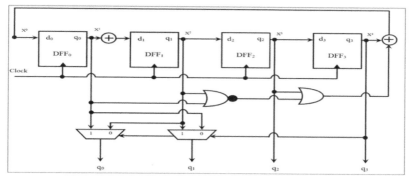

FIGURE 5.3 4-bit BS-CFSR.[8]

In Figure 5.3, a 4-bit BS-CSFR is shown, since $n = 4$ (even number), output 0 is swapped with output 1 using a multiplexer with output 3 as select line for these multipliers. Note that output 3 remains unaltered due to even number of flops. For the proposed design, 8-bit BS-CFSR having second last output bit unchanged is used. It can save transitions that can be equal to $T_{saved} = 2^{(n-1)}$ for each pair of swapped neighboring bits. Since these two bits originally produce 2×2^n transitions, the bit swap will therefore save $T_{saved}\% = (2^{(n-1)})/(2 \times 2^n) = 25\%$.

5.2.2 ORAs

ORA compacts the higher data output rate bits from the CUT into lower outputs and then give to comparator to produce a single pass/fail indication bit. This compaction is done primarily to reduce the test time comparing each bit from CUT with the golden values. There is difference in compaction and compression, compaction is always lossy while latter may or may not be lossy. The only sufficient indicator is the final BIST output which indicates whether circuit has passed or failed.[2] *ORAs using comparator* detect variations in the fault-free and faulty circuits. The expected responses are stored in a ROM called golden response, and a comparator is used to compare bits with the output response analyzer. *Counter-based ORAs* count the number of 0s, 1s, or transitions in the output response of the CUT with the resultant count value of the specific attribute at the completion of the test pattern for giving $1/n$ bit signature as required for the pass/fail indication. The polynomial of the output response of the CUT is divided by the signature analysis and taps of the LFSR are determined by primitive polynomial. The remainder acts as the signature of this division and is then compared to the true schematic already stored in read only memory.

5.3 PROPOSED BIST ARCHITECTURE

The architecture proposed uses BS-CFSR instead of classic LFSR to reduce down the dynamic power consumption in the whole circuit with a penalty of very little area overhead. The CUT is predetermined to be of 4 bits, which here is an RC adder and carry-lookahead (CL) adder. A control signal scan–enabled "SE" is generated by BIST to activate the test mode in various blocks. A selection circuit which is a 2:1 multiplexer isolates primary inputs during test modes from those inputs coming from test pattern generator.

There are two modes, normal and test, in the working of BIST. Normal mode accepts test vectors from primary inputs and feeds it to CUT and produces respective output, while in the test mode, input patterns are generated from BS-CFSR and given to CUT. The output response analyzer is a comparator that compares CUT output with ROM values stored. The synchronization between TPG and ROM is very crucial at this stage for correct response comparison. The response is either pass or fail, pass means BIST is able to sensitize the errors in the CUT, fail means otherwise. A proposed BIST architecture is shown in Figure 5.4. This architecture uses BS-CFSR as pattern generator.

Implementation of Low-Power BIST

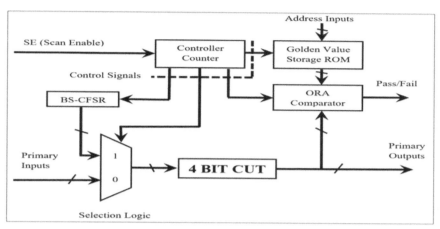

FIGURE 5.4 Proposed BIST blocks.

5.4 RESULTS AND DISCUSSION

RTL analysis has been carried out using industry standard CADENCE tool. A proposed 4-bit RC adder and its successor CL adder are analyzed by using these circuits as CUT. Figure 5.5 shows RTL for the proposed architecture with both RC adder and CL adder.

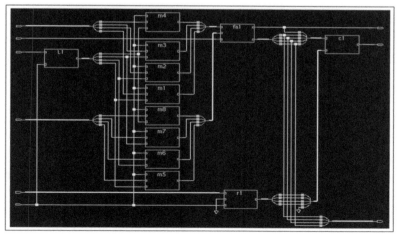

FIGURE 5.5 RTL of 4-bit ripple adder and CL adder.

Timing responses are essential for checking the correctness of the designs made. Fault-free response is collected for both modes—normal mode for first 20 ns and test mode for rest of the time, which is shown in Figure 5.6.

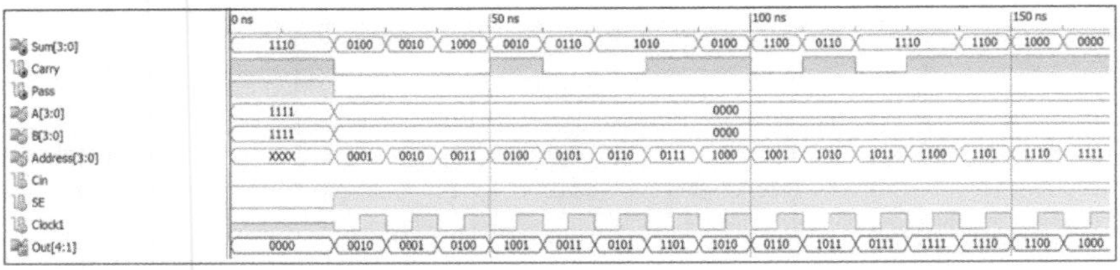

FIGURE 5.6 Timing response for error free CL adder as a CUT.

For normal mode, test vectors were provided from primary inputs "A" and "B," and in test mode, this is done by BS-CFSR, which is then compared with the golden values stored in ROM. The switching between normal and test mode is by a control input "SE." The comparator then produces "Pass" response. If the CUT's output and ROM vectors match, the response is "Pass" = 0, meaning there are no errors in testing.

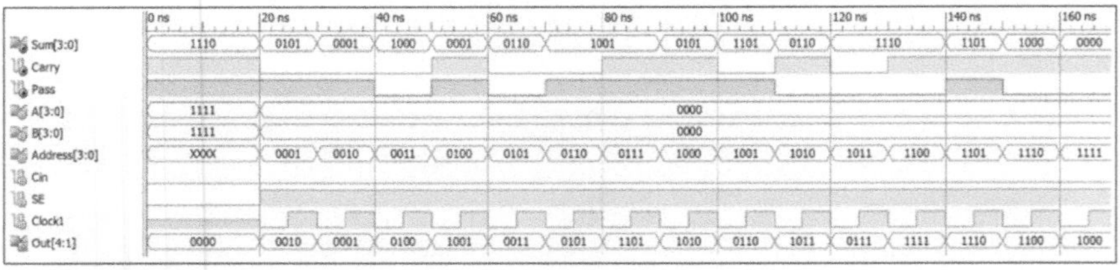

FIGURE 5.7 Timing response with bridging fault at pin A0 of CL adder as a CUT.

If "Pass" = 1, it is otherwise. Now, for the functionality test in the test mode, deliberately a bridging fault is introduced in the CUT at pin A0 bridged with A1. By observing Figure 5.7, essential vectors can be determined to sensitize the same fault. Here, in this case, the essential vectors are XX10 and XX01 from BS-CFSR. The functionality of the CUT is checked by multiple single faults and to sensitize that fault, essential vectors are also explored. RTL compiler is used to evaluate power consumption in various technology nodes such as 45, 90, and 180 nm, each having fast and slow libraries, respectively. One such result analysis is shown in Figure 5.8, giving dynamic and total power of proposed BIST. When compared to classical CFSR, proposed BIST reduces 5.02% of overall power in 45-nm technology with slow library and an overall power of 6.5% in 45-nm technology with fast library.

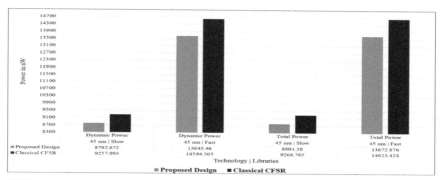

FIGURE 5.8 Dynamic and total power consumption of the proposed work compared with classical CFSR at 45-nm technology.

Critical paths are extremely important when delays are considered for VLSI circuits; timing reports are also generated to check if there are any violations in the design. The worst path delay is 609 and 1982 ps for 45-nm technology as observed with fast and slow libraries, respectively.

5.5 CONCLUSION AND FUTURE WORK

Power consumption is the one of the most essential parts in some IoT devices, and conventional electronic design may not include current testing features. In such cases, the BS-CFSR-based BIST structure provides full fault coverage to sustain the reliability. Therefore, the proposed work, low-power BIST using BS-CFSR for 4-bit RC and CL adder, is designed. It has been found that BIST implemented with CUT as CL adder gave more power reduction that compared to CUT as RC adder. Transition counts in test pattern are reduced by BS-CFSR without any change in the pseudorandom nature. A reduction of 6.5% overall power in the BIST is observed with respect to implemented CL adder. Timing waveforms are also analyzed for both fault-free and faulty circuits. These circuits are studied in normal as well as in test mode. Cadence RTL compiler is used to implement BIST architecture with 45-nm technology node with fast and slow library and does not show any critical path violations. Future work includes implementing the same architecture for higher bit sizes. Another possible scope can be implementing architecture on FPGAs and collect the results to analyze the hardware faults on them and test the functionality.

KEYWORDS

- low-power
- BIST
- BS-CFSR
- CL-adder

REFERENCES

1. Patil, T.; Dhankar, A. In *A Review on Power Optimized TPG Using LP-LFSR for Low Power BIST*, IEEE Sponsored World Conference on Futuristic Trends in Research and Innovation for Social Welfare, 2016; pp 1–4.
2. Charles, E. *A Designer's Guide to Built-in Self-Test*; Springer, 2002; pp 121–132.
3. Abu-Issa, A. S.; Quigley, S. F. Bit-Swapping LFSR for Low-Power BIST. *Electron. Lett.* **2008**, *44*, 1–2.
4. Thirunavukkarasu, V.; Saravanan, R.; Saminadan, V. Performance of Low Power BIST Architecture for UART. *IEEE Int. Conf. Commun. Signal Process.* **2016**, 2290–2293.
5. Wang, S.; Gupta, S. K. DS-LFSR: A BIST TPG for Low Switching Activity. *IEEE Trans. Comput. Aided Des. Integr. Circuits Syst.* **2002**, *21*, 842–851.
6. Kasunde, P.; Shiva Kumar, K. B.; Kurian, M. Z. Improved Design of Low Power TPG Using LPLFSR. *Int. J. Comput. Org. Trends* **2013**, *3*, 102–106.
7. Swapna, S.; Sunilkumar, S. M. Review of LP-TPG Using LP-LFSR for Switching Activities. *Int. J. Adv. Res. Comput. Sci. Softw. Eng.* **2015**, *5* (2), 1–5.
8. Trivedi, R.; Dhariwal, S.; Kumar, A. Comparison of Various ATPG Techniques to Determine Optimal BIST. *IEEE Int. Conf. Intell. Circuits Syst.* **2018**; 93–98.
9. Dhariwal, S.; Trivedi, R. Design and Analysis of Power and Area Efficient Novel Concurrent Cellular Automation Logic Block Observer BIST Structure. *Int. J. Performability Eng.* **2020**, *16*, 19–26.

CHAPTER 6

A Review of a Low-Power CMOS Comparator

TEJENDER SINGH[1,2*] and SUMAN LATHA TRIPATHI[1*]

[1]*School of Electronics and Electrical Engineering, Lovely Professional University, Punjab, India*

[2]*CMR Institute of Technology, Hyderabad, Telangana, India*

*Corresponding author. E-mail: tejendersingh27@gmail.com; suman.21067@lpu.co.in

ABSTRACT

In last decade, there was tremendous progress in smartphones where it became more like companion, health checking, performance monitoring, and anticipating one's daily activities. A major part of it was evolved by sensor technology. A proximity sensor is one of the most used devices in commercial and industrial sectors. It is a device that detects the signal if any object comes physically close to the sensor. The present mobile phones use IR-based proximity sensors that detect the presence of cheek and human ear. The main block of the proximity sensor is analog–digital converter. This chapter gives a detail analysis of designing various complementary metal oxide semiconductor (CMOS) comparators that reduce the power consumption and use low voltage and detect the presence of signal at high speed. A review is done related to different comparators where a comparison is done between different models of comparators, and it is observed that single-tail current dynamic latch comparator has less power and less power delay product. Finally, the most efficient comparator from the discussed comparators can be made compatible with wide range of Internet of things applications such as health care, environmental monitoring, and manufacturing unit.

6.1 INTRODUCTION

The fast-evolving Internet of things (IOT) industry strives hard for high-speed and low-power devices. In IoT enabled system, comparators play a wide role, which reduces the size of the system and costs.[16] It uses a sensor to collect the data and transmit it over an Internet for the applications in factories, wearable devices, mobile phones, etc. considering an application of mobile phone proximity sensor plays an important role. A proximity sensor is a type of electronic sensor which detects the presence of objects within its area without any physical contact. Since 1983, Fargo controls proximity sensors are made to meet the industrial and commercial requirements which are of the highest quality, repeatability, and durability. Today's smartphones are packed with almost 14 sensors that bring intelligence and awareness, and they also produce raw data on motion, environmental conditions, and location of us.[1]

Proximity sensor comprises an infrared (IR) light detector, infrared LED, LED driver, and analog–digital converter (ADC). It detects the outside object based on the closeness of an object to our ear and turn off the screen to avoid inadvertent touches by the cheek. This reduces the display power consumption. The basic building blocks of ADC are voltage reference, op–amp and comparator. An operational amplifier-based comparator is shown later in Figure 6.1.

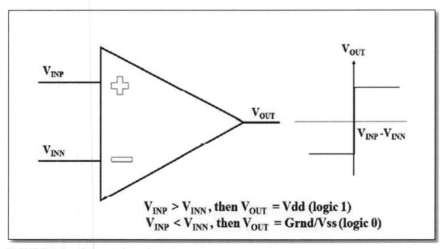

FIGURE 6.1 Op-amp-based comparator.

A Review of a Low-Power CMOS Comparator

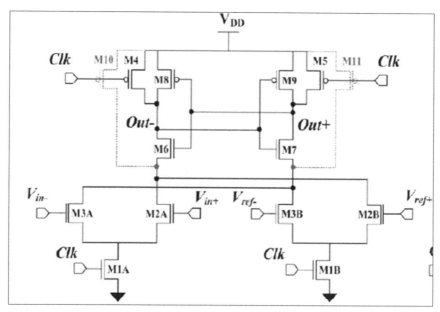

FIGURE 6.2 Schematic view of differential current-sensing amplifier-based comparator.

A differential current-sensing amplifier-based comparator is shown in the previous Figure 6.2, and it is one of the most used complementary metal oxide semiconductor (CMOS) devices that consume less power, high speed, high input impedance, and rail-to-rail swing as output similar to dynamic latch comparators. One of the main design parameters is offset voltage that should be less. The accuracy of the dynamic latch is decreased with the help of device-generated offset voltages and random noise. A preamplifier is used prior to regenerative stage in order to decrease an offset voltage that amplifies the small input–large output signal and reduces a kickback noise.[2] The main drawback of preamplifier stage is the large static power dissipation and without using preamplifier stage. It can be used for low-power application devices that offer a high speed.

Latch-type dynamic comparator is extensively used in mixed-signal circuits and analog circuits that generate a binary output by comparing two analog input voltages. Many authors have attempted and presented architectural modification in the ***dynamic latch comparator*** to handle and optimize various parameters like ***speed, power, low-voltage operation,*** offset voltage minimization, kickback noise reduction, etc. Due to its power efficiency, fast decision-making ability, and low supply necessity, a dynamic latch comparator with positive feedback is currently the most used architecture in high-speed ADCs.

There are classically two types of the dynamic latch-based comparators, namely, **single-tail current (STDLC) and double-tail current (DTDLC)**. A double-tail comparator has better optimization for balancing the speed, offset voltages, and power by adding a degree of freedom by extending the input common mode range compared to a STDLC comparator and it also reduces the kickback noise via providing isolation among the input and output which can be operated at low supply voltages. A little modification is done to the present double-tail comparator by adding few transistors to it which is of high speed, area optimization, and low-power consumption.[3]

6.2 COMPARISION OF DIFFERENT COMPARATORS

In this chapter, we are going to compare four different comparators—STDLC, DTDLC, modified DTDLC (MDTDLC), double tail dynamic latch comparator with clock (DTDLC-CLK).

6.2.1 SINGLE-TAIL CURRENT DYNAMIC LATCH COMPARATOR (STDLC)

One of the most used ADCs that provide high input impedance, zero static power consumption, and output swing is STDLC. Its operation is classified into dual phase, they are reset phase and regeneration phase.

Here, clock is kept low in reset phase, and M_{tail} transistor is maintained in OFF state. A logical level V_{dd} is obtained by pulling both the output terminal Out_n and Out_p by M_7 and M_8 transistors and is called as starting state of regeneration phase. When CLK switch to V_{dd}, the M_{tail} transistor will turn ON, while M_7 and M_8 will go to OFF state by switching back the circuit to regeneration phase, whereas terminals Out_n and Out_p will get discharged, which were precharged at the first stage due to supply. The input voltage applied at terminals, that is, INN and INP gives the discharge rate as shown in the Figure 6.3.

Assuming that $V_{INP} > V_{INN}$ such that Out_n discharges faster than Out_p and the Out_n terminal falls down to $V_{dd}-V_{tp}$ such that PMOS transistor

A Review of a Low-Power CMOS Comparator

(M_6) turns ON and starts the second stage, that is, regeneration process produced by cross-coupled inverter by making the output terminal Out_n to completely discharge to GND and to charge the output terminal Out_p to supply voltage V_{dd}.[4-6]

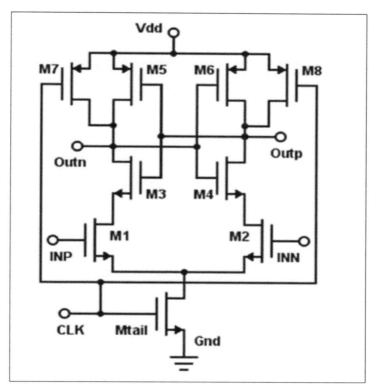

FIGURE 6.3 Schematic view of single-tail current dynamic latch comparator.

Analysis:

While assumption of the comparator delay is divided into two parts, that is, t_0 and t_{latch}, where t_0 is the discharge time period due to load capacitance (C_L), when transistor PMOS is switched ON, an output terminal Out_n discharges faster due to transistor M_1.

$$t_{latch} = \frac{C_L \cdot |V_{thp}|}{I_1} \cong \frac{C_L \cdot |V_{thp}|}{I_{tail}} \qquad (6.1)$$

where $I_1 = I_{tail}/2$

$$I_{in} = \left(\frac{I_{tail}}{2}\right) + \left(\frac{g_{m1,2}}{V_{in}}\right)$$

ΔV_{in} is the input differential voltage, for small change in ΔV_{in}, the drain current I_1 is moved toward constant, and it is half of the tail current. A t_{latch} is a total latch current due to dual inverters which are cross coupled. The output is equal to 0.5 of a supply rail voltage ($\Delta V_{out} = V_{dd}/2$). A comparator followed by latch enhances the voltage to swing fully rail to rail. The latching time is expressed by

$$t_{latch} = \frac{C_L}{g_{m(eff)}} \cdot ln\left(\frac{\Delta V_{out}}{\Delta V_0}\right) \cong \frac{C_L}{g_{m(eff)}} \cdot ln\left(\frac{V_{dd}/2}{\Delta V_0}\right) \quad (6.2)$$

A transconductance of cross-coupled inverter is indicated by $g_{m(eff)}$ by an eq 6.1, the ΔV_0 initial voltage difference at the input and is given as

$$\Delta V_0 = \left|V_{Out_n}(t=t_0) - V_{Out_p}(t=t_0)\right|$$

$$= \left|V_{thp}\right| - \frac{I_2 \cdot t_0}{C_L} \quad (6.3)$$

$$\Delta V_0 = \left|V_{thp}\right|\left(1 - \frac{I_2}{I_1}\right)$$

The input current difference ΔI_{in} is due to the individual current I_1 and I_2, which is much small. Therefore, the equation can be approached as ($I_{tail}/2$). So eq 6.3 becomes

$$\Delta V_0 = \left|V_{thp}\right| \Delta\left(\frac{\Delta I_{in}}{I_1}\right)$$

$$\approx 2.\left|V_{thp}\right| \cdot \left(\frac{\Delta I_{in}}{I_{tail}}\right)$$

$$= 2.\left|V_{thp}\right| \cdot \frac{\sqrt{\beta_{1,2} \cdot I_{tail}}}{I_{tail}} \cdot \Delta V_{in}$$

Finally,

$$\Delta V_0 = 2 \cdot |V_{thp}| \cdot \frac{\sqrt{\beta_{1,2}}}{I_{tail}} \cdot \Delta V_{in} \quad (6.4)$$

The factor $\beta_{1,2}$ is input transistor current now replacing ΔV_0 in eq 6.2 with the t_0 value from eq 6.1 and the t_0, that is, total delay can be calculated as

$$t_{delay} = t_0 + t_{latch}$$

$$t_{delay} = 2 \cdot \frac{C_L \cdot |V_{thp}|}{I_1} + \frac{C_L}{g_{m(eff)}} \cdot ln\left(\frac{V_{dd}}{4 \cdot |V_{thp}| \Delta V_{in}} \cdot \sqrt{\frac{I_{tail}}{\beta_{1,2}}}\right) \quad (6.5)$$

6.2.2 DOUBLE-TAIL CURRENT DYNAMIC LATCH COMPARATOR (DTDLC)

This comparator circuit is made of less number of transistors and has two stages of different tail currents. In stage 1, it produces low offset by handling a small current, whereas in stage 2, it provides short delay time by handling a large current, and the circuit is made up of latch as shown in the Figure 6.4.

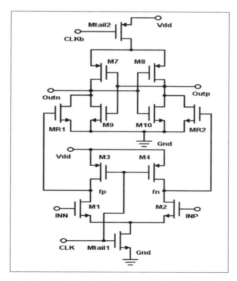

FIGURE 6.4 Schematic view of double-tail current dynamic latch comparator.

This circuit at differential stage optimizes very less current to achieve longer integration time and large-tail current for fast latching. It works in two phases similar to STDLC, that is, reset phase and regeneration phase. In phase 1, that is, reset stage (clock is low) and the current sources M_{tail1} and M_{tail2} are switched OFF and result in zero static power consumption.[6] The terminal fp and fn will be pulled toward supply voltage V_{dd} by M_3 and M_4 transistors that are in ON state. Similarly, Out_p and Out_n are pulled down to the GND by MR1 and MR2 transistors that are switched ON. In the second stage, clock is at high where current sources are turned ON by making transistor M_3 and M_4 to switch OFF. Therefore, fn and fp with different speeds get discharged, and when any one of the outputs of reset stage reaches below V_{th} of MR1, MR2 transistors turn OFF. In a latch, there are two outputs due to the positive feedback, one reaches to V_{dd} and other reaches to ground. The amplified differential input voltage is driven to cross-coupled inverters by the middle stage and reduces kickback noise.[7–12]

Analysis:

This has two delays, that is, t_0 and t_{delay}. The t_0 delay is due to the discharge of load capacitance C_L until the M_9 or M_{10} NMOS transistor gets turn ON and therefore t_0 is calculated as

$$t_0 = \frac{V_{thn} \cdot C_L}{I_{B1}} = 2 \cdot \frac{V_{thn} \cdot C_L}{I_{tail2}} \tag{6.6}$$

Regeneration phase starts after transistor NMOS (M_9 or M_{10}) gets turned ON. Here, in the previous equation, the M_9 transistor produces the drain current I_{B1} that is approximately equal to $I_{tail}/2$. When the particular output Out_n gets discharged to ground, it makes the transistor M_8 to turn ON and it charges toward the power supply (V_{dd}). The initial voltage difference is defined as follows:

$$\Delta V_0 = \left| V_{Out_p}(t=t_0) - V_{Out_n}(t=t_0) \right|$$

$$= \left| V_{thn} \right| - \frac{I_{B2} \cdot t_0}{C_L}$$

$$\Delta V_0 = \left| V_{thn} \right| \left(1 - \frac{I_{B2}}{I_{B1}} \right) \tag{6.7}$$

A Review of a Low-Power CMOS Comparator

In the following equation, I_{B1} is the left side branch current, I_{B2} is the right side branch current of regeneration stage, and the latch current is given by $\Delta I_{latch} = |I_{B1} - I_{B2}| = g_{mR1,2} \Delta V_{fn/fp}$

$$\Delta V_0 = |V_{thn}| \cdot \left(\frac{\Delta I_{latch}}{I_{B1}}\right) \approx 2 \cdot |V_{thn}| \cdot \left(\frac{\Delta I_{latch}}{I_{tail2}}\right)$$

$$\Delta V_0 = 2 \cdot |V_{thn}| \cdot \left(\frac{g_{mR1,2}}{I_{tail2}}\right) \cdot \Delta V_{fn/fp} \tag{6.8}$$

The differential voltage among fn and fp is $\Delta V_{fn/fp}$ at a time t_0 as shown in the previous eq 6.8, which is calculated at the first stage, and at transferring stage, transconductance is given by $g_{mR1,2}$ which is made up of MR1 and MR2 transistors. To improve the design, the two important parameters are ΔV_0 and regeneration time. At intermediate stage, amplified differential voltage causes the latch to imbalance.

$$\Delta V_{fn/fp} = |V_{fn}(t = t_0) - V_{fp}(t = t_0)|$$

$$= t_0 \cdot \frac{I_{N1} - I_{N2}}{C_{L,fn(fp)}}$$

$$\Delta V_{fn/fp} = t_0 \cdot \frac{g_{m1,2} \cdot \Delta V_{in}}{C_{L,fn(fp)}} \tag{6.9}$$

Here, I_{N1} and I_{N2} represents discharging currents of M_1 and M_2 transistor where difference in currents depend on the differential input voltage given by ($\Delta I_{IN} = g_{m1,2} \Delta V_{in}$). Equation 6.4 becomes,

$$\Delta V_0 = \left(\frac{2V_{thn}}{I_{tail2}}\right)^2 \cdot \frac{C_L}{C_{L,fn(fp)}} \cdot g_{mR1,2} \cdot g_{m1,2} \cdot \Delta V_{in} \tag{6.10}$$

Therefore, T_{delay} is total delay and can be calculated by substituting the previous eq 6.10 in latch regeneration time.

$$t_{delay} = 2 \cdot \frac{V_{thn} \cdot C_L}{I_{tail2}} \cdot \frac{C_L}{g_{m(eff)}} \cdot \ln\left(\frac{V_{dd} \cdot I^2_{tail2} \cdot C_{L,fn(fp)}}{8 \cdot V^2_{thn} \cdot C_L \cdot g_{mR1,2} \cdot g_{m1,2} \cdot \Delta V_{in}}\right) \tag{6.11}$$

6.2.3 MODIFIED DOUBLE-TAIL CURRENT DYNAMIC LATCH COMPARATOR (MDTDLC)

An improved version of DTDLC is shown in Figure 6.5, which improves the delay. The modified architecture increases the $\Delta V_{fn/fp}$ voltage and the initial difference voltage ΔV_0 and reduces the latch delay, thereby decreasing the overall delay. We have added two extra PMOS transistor in the first stage MC1 and MC2, that is, control transistors that are set parallel to reset switch transistors M_3 and M_4 connected in a cross-coupled method. Its process is divided into two phases when clock is zero; both M_{tail1} and M_{tail2} transistors remain OFF state and avoid static power. Transistor M_3 and M_4 both will be pulled toward V_{dd} and control signals MC1 and MC2 will remain in cut-off state by making the outputs Out_n and Out_p discharged to GND with the help of MR1 and MR2. When the clock signal is high, that is, CLK = V_{dd}, the circuit enters into the regeneration phase, the tail transistors get ON, and switch transistors M_3 and M_4 and the control transistor will be in switch OFF state.[13]

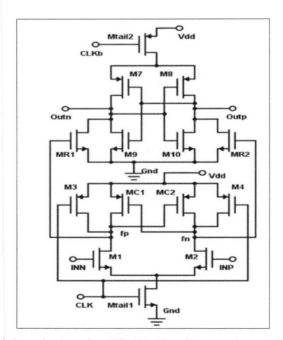

FIGURE 6.5 Schematic view of modified double-tail current dynamic latch comparator.

The two terminals *fn* and *fp* are near to V_{dd}. The voltage will start discharging at different rates based on the applied input differential voltage. By assuming ($V_{INN} > V_{INP}$) discharging of *fp* is quicker than *fn*, the PMOS control transistor turns ON and pull *fn* to V_{dd} and control transistor MC1 gets turn OFF and makes *fp* to discharge completely. An exponential increase in the differential voltage $\Delta V_{fn/fp}$ reduces the regeneration latch time.

The analysis for delay expression is similar to the DTDLC; here t_0 is the delay time after which latch generation starts until the NMOS transistor of cross-coupled inerter is turned ON and pulls anyone of the Out_p or Out_n terminals and regeneration starts. Here, t_0 is derived as

$$\Delta V_0 = V_{thn} \cdot \frac{\Delta I_{latch}}{I_{B1}} \approx 2 \cdot V_{thn} \cdot \frac{\Delta I_{latch}}{I_{tail2}}$$

$$\Delta V_0 = 2 \cdot V_{thn} \cdot \frac{g_{mR1,2}}{I_{tail2}} \cdot \Delta V_{fn/fp} \qquad (6.12)$$

Now the $\Delta V_{fn/fp}$ at $t = t_0$

$$\Delta V_{fn/fp} = \Delta V_{fn/fp(0)} \cdot exp^{(A_v-1) \cdot t/\tau} \qquad (6.13)$$

And here
$$\frac{\tau}{A_v - 1} \cong \frac{C_{L,fn(fp)}}{G_{m(eff1)}}$$

and

$$\Delta V_{fn/fp(0)} = 2 \cdot |V_{thp}| \cdot \frac{g_{m1,2} \cdot \Delta V_{in}}{I_{tail1}} \qquad (6.14)$$

Finally, ΔV_0 can be expressed as the following:

$$\Delta V_{(0)} = 4 \cdot V_{thn} \cdot |V_{thp}| \cdot \frac{g_{mR1,2}}{I_{tail2}} \cdot \frac{g_{m1,2} \cdot \Delta V_{in}}{I_{tail1}} \cdot exp\left(\frac{G_{m(eff1)} \Delta t_0}{C_{L,fn(fp)}}\right) \qquad (6.15)$$

By using eqs 6.5, 6.6, and 6.15, total delay can be expressed as the following:

$$t_{delay} = 2 \cdot \frac{V_{thn} \cdot C_L}{I_{tail2}} \cdot \frac{C_L}{g_{m(eff)} + g_{m1,2}} \cdot ln\left(\frac{V_{dd}/2}{4 \cdot V_{thn} \cdot |V_{thp}| \cdot (g_{mR1,2}/I_{tail2}) \cdot (g_{m1,2} \cdot \Delta V_{in}/I_{tail1}) \cdot exp\left((G_{m(eff1)} t_0 / C_{L,fn(fp)})\right)}\right) \qquad (6.16)$$

6.2.4 TWO-STAGE DYNAMIC COMPARATOR WITHOUT INVERTED LOCK (DTDLC-CLK)

This version of a **DTDLC-CLK** consists of two clock signals CLK and CLKb (CLK′), and it has resolved a problem of synchronization between them by CLK′ with Di nodes, and by this CLK load has been lessened, input offset voltage has also been reduced. Architecture of this is similar to DTDLC as shown in Figure 6.6. A gain is obtained as the output latch stage from both the input NMOS transistors M_{10} and M_{11} and PMOS transistors M_{12} and M_{13}. The improved offset has a trade-off with increased delay and a weakened output load. This is because of M_{12} and M_{13} transistors which use Di node voltage as their CLK′ signal. The tail current M_{tail2} is divided to half and is separated in two parts by M_{12} and M_{13} transistors.[14,15]

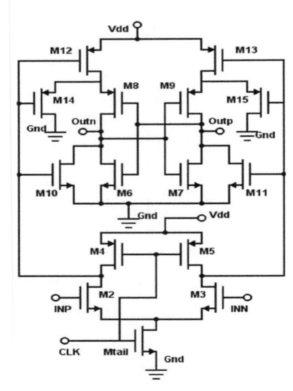

FIGURE 6.6 Schematic view of double-tail current dynamic comparator without inverted clock.

The transient behavior is shown by two delay parts, that is, total delay t_0 and total latch time t_{latch}. The capacitive discharge of load capacitance C_L is calculated by total delay time t_0 and it is given as:

$$t_0 \equiv \frac{V_{thn} \cdot C_L}{I_{B1 \, or \, B2}} \approx \frac{V_{thn} \cdot C_L}{I_{MT12(MT13)}} \tag{6.17}$$

Now, the first NMOS transistor M_6 or M_7 gets turn ON and regeneration starts. The current through the transistor M_6 or M_7 is I_{B1} or I_{B2}, which is replaced by the current I_{MT12} or I_{MT13} of transistor MT12 or MT13. Once the transistor M_6 or M_7 is turned ON, the output terminal Out_n will start discharging to GND and make the transistor on latch side to turn ON state, and it charges the output terminal to V_{dd}, that is, toward supply voltage. The regeneration time is given by

$$\Delta V_0 = \left| V_{Out_p}(t=t_0) - V_{Out_n}(t=t_0) \right| \tag{6.18}$$

The previous equation is simplified and it is given as follows:

$$\Delta V_0 = |V_{thn}| - \frac{I_{B2} \cdot t_0}{C_L}$$

$$\Delta V_0 = |V_{thn}| - \frac{I_{B2}}{C_L} \cdot \frac{C_L \cdot V_{thn}}{I_{B1}} \tag{6.19}$$

The simplified expression is

$$\Delta V_0 = V_{thn} \cdot \left(1 - \frac{I_{B2}}{I_{B1}}\right) = V_{thn} \cdot \left(\frac{\Delta I_{latch}}{I_{B1}}\right) \tag{6.20}$$

$$\Delta V_0 = V_{thn} \cdot \frac{\Delta I_{latch}}{I_{B1}}$$

$$\approx V_{thn} \cdot \left(\frac{\Delta I_{latch}}{I_{MT12(MT13)}}\right)$$

$$\Delta V_0 = V_{thn} \cdot \left(\frac{g_{mR1,2}}{I_{MT12(MT13)}}\right) \cdot \Delta V_{fn/fp} \tag{6.21}$$

In the previous eq 6.21, at time t_0 during initial stage, a differential voltage is calculated between fn and fp which is given by $\Delta V_{fn/fp}$, and $g_{mR1,2}$ is transconductance at transfer stage which is made up of M_{10} and M_{11} transistors.

$$\Delta V_{fn/fp} = \left| V_{fn}(t=t_0) - V_{fp}(t=t_0) \right|$$

$$= t_0 \cdot \frac{I_{N1} - I_{N2}}{C_{L,fn(fp)}} \quad (6.22)$$

$$\Delta V_{fn/fp} = t_0 \cdot \frac{g_{m1,2} \cdot \Delta V_{in}}{C_{L,fn(fp)}}$$

Substituting eq 6.22 in 6.21, ΔV_0 can be expressed as eq 6.23

$$\Delta V_0 = \left(\frac{V_{thn}}{I_{MT12(MT13)}}\right)^2 \cdot \frac{C_L}{C_{L,fn(fp)}} \cdot g_{mR1,2} \cdot g_{m1,2} \cdot \Delta V_{in} \quad (6.23)$$

The total delay can be calculated by substituting the value of t_0 and ΔV_0 in latch regeneration time equation

$$t_{delay} = t_0 + t_{latch} = 2 \cdot \frac{V_{thn} \cdot C_L}{I_{tail2}} + \frac{C_L}{g_{m(eff)}} \cdot \ln\left(\frac{V_{dd}/2}{\Delta V_0}\right) \quad (6.24)$$

Finally, the delay expression for the two-stage dynamic comparator (DTDLC-CLK) is expressed by the equation

$$t_{delay} = \frac{V_{thn} \cdot C_L}{I_{MT12(MT13)}} + \frac{C_L}{g_{m(eff)}} \cdot \ln\left(\frac{V_{dd} \cdot (I_{MT12(MT13)})^2 \cdot C_{L,fn(fp)}}{V_{thn}^2 \cdot C_L \cdot g_{mR1,2} \cdot g_{m1,2} \cdot \Delta V_{in}}\right) \quad (6.25)$$

6.3 RESULT ANALYSIS

Comparison of different comparator architectures based on delay, power, and power delay product is shown in Table 6.1.

TABLE 6.1 Comparisons of Different Comparator Architectures.

Comparator architecture	No of transistors	Delay (ps)	Power (µw)	Power delay product (fJ)
STDLC	9	77.70	26.89	2.09
DTDLC	14	66.40	52.40	3.48
MDTDLC	16	54.50	147.70	8.05
DTDLC-CLK	17	75.40	57.20	4.31

The prelayout simulated results are calculated where the delay is calculated at 50% of CLK rising edge, and the difference between two outputs is $V_{dd}/2$ half of the input supply voltage. From the previous table, we can tell that the delay of STDLC is 77.70 ps and power consumption is 26.89 µw, and therefore, the power delay product is 2.09 (fJ). This is less comparatively from all as shown in Figure 6.7

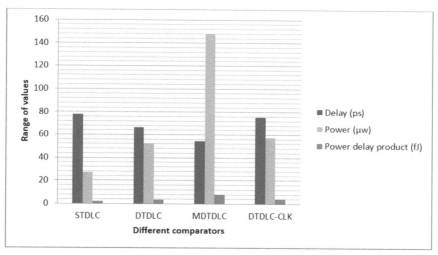

FIGURE 6.7 Comparisons of different comparator.

6.4 CONCLUSION

This chapter gives a detailed analysis that has been done on different conventional dynamic latch comparators that are studied based on their performance parameters such as delay, power, and power delay product and found that power delay product of STDLC is 2.09 (fJ) in comparison to DTDLC, MDTDLC, and DTDLC-CLK. So, we can use STDLC for various low power applications instead of other comparators. It also describes the functionality and design of comparators that can be used in any of the IOT-enabled systems and can be integrated to the devices for day-to-day usages like smart phones, traffic control, agriculture monitoring, healthcare monitoring, cybercrime detection, and many more.

KEYWORDS

- comparator
- analog–digital converter
- low power consumption
- offset voltage
- high speed
- power delay product

REFERENCES

1. Chan, C. H.; Zhu, Y.; Chio, U. F.; Sin, S. W.; Pan, U. F.; Martins, R. P. In *A Reconfigurable Low-Noise Dynamic Comparator with Offset Calibration in 90nm CMOS,* IEEE Asian Solid State Circuits Conference, A-SSCC, 2011; pp 233–236.
2. Katyal, V.; Geiger, R. L.; Chen, D. J. In *A New High Precision Low Offset Dynamic Comparator for High Resolution High Speed ADCs,* Asia Pacific Conference on Circuits and Systems, APCCAS, IEEE Proceeding, 2006; pp 5–8.
3. He, J.; Zhan, S.; Chen, D.; Geiger, R. In *A Simple and Accurate Method to Predict Offset Voltage in Dynamic Comparators,* International Symposium on Circuits and Systems, ISCAS, IEEE Proceeding, 2008; pp 1934–1937.
4. Allen, P. E.; Holberg, D. R. *CMOS Analog Circuit Design;* Oxford University Press: Oxford, 2002.
5. Uthaichana.; Patheera.; Leelarasmee, E. In *Low Power CMOS Dynamic Latch Comparators,* Conference on Convergent Technologies for the Asia-Pacific Region, TENCON, IEEE Proceeding, 2003; pp 605–608.
6. Gandhi, P. P.; Devashrayee, N. M. A Novel Low Offset Low Power CMOS Dynamic Comparator. *Analog Integr. Circuits Signal Process.* **2018,** *96,* 147–158.
7. Schinkel, D.; Mensink, E.; Klumperink, E.; van Tuijl, E.; Nauta, B. In *A Low Offset Double Tail Latch Type Voltage Sense Amplifier,* 18th Annual Workshop on Circuits, Systems and Signal Processing, Pro RISC; Technology Foundation STW: Utrecht, 2007; pp 89–94.
8. Vemu, S. R.; Mowlika, P. S. S. N.; Adinarayana, S. An Energy Efficient and High Speed Double Tail Comparator Using Cadence EDA Tools. *IEEE Proc.* **2017.**
9. Savani, V.; Devashrayee, N. M. Analysis and Design of Low Voltage Low Power High Speed Double Tail Current Dynamic Latch Comparator. *Analog Integr. Circuits Signal Process., Springer* **2017,** *93,* 287–298.
10. Savani, V.; Devashrayee, N. M. Analysis of Power for Double-Tail Current Dynamic Latch Comparator. *Analog Integr. Circuits Signal Process., Springer* **2019,** *100,* 345–355.

11. Avaneesh Dubey, K.; Nagaria, R. K. Low-Power High-Speed CMOS Double Tail Dynamic Comparator Using Self-Biased Amplification Stage and Novel Latch Stage. *Analog Integr. Circuits Signal Process., Springer* **2019**, *101*, 307–317.
12. Samaneh, B.-M.; Moghaddam, M. S. In *Analysis and Design of Dynamic Comparators in Ultra-low Supply Voltages*, 22nd Iranian Conference on Electrical Engineering, ICEE, IEEE Proceeding, 2014; pp 255–258.
13. Bernhard, W. Current Sense Amplifiers for Embedded SRAM in High Performance System on a Chip Designs. *Springer Sci. Bus. Media* **2013**, *12*.
14. Jeon, H. J.; Kim, Y.-B. In *A CMOS Low Power Low Offset and High Speed Fully Dynamic Latched Comparator*, International SoC Conference, SoCC, IEEE Proceeding, 2010; pp 285–288.
15. van Michel, E.; van Tuijl, E.; Geraedts, P.; Schinkel, D.; Klumperink, E.; Nauta, B. In *A 1.9 µW 4.4fJ/Conversion-Step 10b 1MS/s Charger Distribution ADC*, Solid-State Circuits Conference, ISSCC 2008, Digest of Technical Papers. IEEE International, 2008; pp 244–610.
16. Ali Hassan, H.; Hassan Mostafa, Y.; Khaled Salama, Z. N.; Ahmed Soliman, M. A Low-Power Time-Domain Comparator for IoT Applications. *IEEE Proc.* **2018**. 978-1-5386-7392-8/18/$31.00.

CHAPTER 7

Encryption and Decryption Algorithms for IoT Device Communication

ANANYA DASTIDAR* and SONALI MISHRA

Department of Instrumentation and Electronics Engineering, College of Engineering and Technology, Bhubaneswar 751029, Odisha, India

*Corresponding author. E-mail: adastidar@cet.edu.in

ABSTRACT

The current trend in computing and data communication is the Internet of things (IoT) that makes use of an interconnected network of devices that utilizes cloud servers for storage and analysis of data. As the number of devices increases progressively and with the reduction in human intervention in the device control and management, different challenging security issues are being faced in data communication. Moreover, the internetworking of these devices and the use of cloud servers make the devices vulnerable to various types of threats. Thus, data safety and security is a challenge while implementing a device network over the Internet. Cryptography, watermarking, and steganography are a few methods that are employed by software engineers that help us to prevent data theft or corruption to a large extent. Image encryption is a method that sends out the data in a secure manner over the network so that any illegal decryption of the message is prevented. This chapter deals with various encryption–decryption algorithms that can be implemented during the communication taking place in an IoT network.

7.1 INTRODUCTION

The exponential increase in data traffic poses a challenge for the communication infrastructure as we move over to the fifth-generation wireless networks from the current fourth-generation standards. The World Wide Web is used as a medium to circulate a huge amount of digital information throughout the world every day. Cryptography deals with the storage and transmission of information in a meticulous method so that only the authentic owner (sender/receiver) of the same can read and process it. Data security and data anticorruption are two areas of cryptography that is under continuous research and development. The advancement of encryption is headed for a future containing boundless risks and opportunities. Nowadays, encryption is vital as it permits the user to protect their data and maintain its security as an unauthorized user will be unable to access it.

Internet of devices that is widely referred to as the Internet of things (IoT) is a device network capable of sensing and gathering data and thus sending it to the Internet consisting of sensors that are able to interface to the cloud by the Internet. The data that uses cloud servers for storage can be retrieved by some analysis for onward transmission to a desired user. IoT is the link between the physical world and the cyberspace where the networking, connectivity speed, and the protocols used usually rely on particular IoT applications deployed over Web-based network. Owing to the necessity of communication over the device network, maintaining data security is an important issue. The domains of transportation, smart grid, sports, patient monitoring, and automobile sensor networks are all going to be ruled by IoT.[1] This chapter discusses various techniques of data encryption and decryption that can be used for the trending IoT device communications.

7.2 DATA SECURITY FEATURES IN IOT DEVICE COMMUNICATION

Security in an IoT network is necessitated in at least three areas of the IoT system—the security of the device, security in interdevice data communication, and the security of the stored data. Some of the important features are discretion or confidentiality, reliability or integrity, validation or authentication, and nonrepudiation.[2] One of the features associated with security in data communication between IoT devices is confidentiality wherein the data to be exchanged between devices must be private to the sender and receiver and an attacker must not gain access to the same. Another feature is the data integrity where the reliability of the data is maintained. Authentication is another feature where the origin of the message must be verifiable. Nonrepudiation is another feature where the owner cannot deny the ownership of the data which is important for maintaining security. Research in IoT device communication security has garnered a lot of interest in recent years with the solutions being provided not only for layer security but also for end-to-end security.[3] There are different methods by which security in IoT device communication may be implemented like cryptography, watermarking, and steganography. While cryptography can be subdivided into secret (private) and open (public) key algorithms which can be used for IoT, watermarking can be subdivided into spatial and frequency domain algorithms and steganography can be subdivided into technical and linguistic algorithms. The subsequent sections will discuss essentially about the cryptography algorithms that have found the use for implementing security in IoT device communication.

7.3 CRYPTOGRAPHY

Cryptography enables securing the data in a cryptosystem, the latter being a combination of input data (plaintext), encryption algorithm, encryption keys, encrypted data (ciphertext), and decryption algorithm and decryption keys that are used to recover the input data. Cryptography is one of the best methods for enhancing the security of communication channel in IoT device network.[4]

The encoding of the input data or information in order to convert it to a scrambled (cipher) text is known as encryption and it is associated with an encryption key (sender) such that it can only be accessed by an authorized user. The decryption is the counter process of encryption where the secret message (or ciphertext) is converted back to the unique plaintext using a decryption key (receiver). The basic model of cryptography includes the process how the input data is encrypted followed by the generation of ciphertext or secret text and finally the output data is obtained following the decryption process.[5] There are two types of keys (symmetric and asymmetric) used which are carried out by encryption–decryption.[6]

The symmetric-key/private-key algorithm is where only the communication parties must have the identical keys in order to facilitate the proper link between the devices.[5] In comparison to asymmetric-key algorithms, symmetric key algorithms are more secure and faster and require lesser power as compared to the former.

The other type of key is the asymmetric key/public key where only the encryption key is available to everyone to use but the decryption key is not and only the registered user (receiver) has the correct decryption key that allows the message to be read after decryption.[5]

7.3.1 DIFFERENT CRYPTOGRAPHY ALGORITHMS

7.3.1.1 RIVEST–SHAMIR–ADLEMAN

This algorithm is a symmetric (public) key algorithm for cryptography that provides excellent safety in the IoT and Message Queuing Telemetry Transport systems.[7,8] A key generator is used by the RSA (Rivest–Shamir–Adleman) algorithm that provides two numbers (primes) representing the two varieties of keys that are used in the encryption–decryption process. This type of cryptography is used for electronic communications and online transactions but it faces the challenge of computational complexity and cost inefficiency. For guaranteed security, a key of more than 2048 bits is essential, but the use of such large key size makes it unsuitable for IoT applications that use small-sized devices, wireless communication, and fast processing.

7.3.1.2 ELLIPTIC-CURVE CRYPTOGRAPHY

The use of ECC (elliptic-curve cryptography) has been found to be prominent in IoT[9] which is another example of public-key algorithm. The whole algorithm is based on three steps—first is the key generation (public and private) where the receiver's public key finds use for message encryption by the sender while for the decryption of the message, receiver's private key is used. While ECC encryption is able to offer an improved level of security owing to its use of a shorter length key as compared to RSA, yet owing to its complexity and cost ineffectiveness, it is not that popular in IoT device communication. It has been reported that ECC algorithm offers enhanced performance in terms of memory, energy and size of key, and time for execution over RSA, while the latter outperforms ECC in verification and encryption.[10]

7.3.1.3 SECURE FORCE

The various steps in the secure force algorithm include key generation followed by key management and encryption and decryption processes. The SF (secure force) algorithm[6] begins with key expansion and offers a reduced probability of a fragile key and enhanced strength of key by employing the process of expansion of key and selection of round key. The 64-bit cipher key is partitioned into four equal parts with 2^4 bits per part with every bit organized horizontally as well as vertically as a 4 × 4 matrix over which left-shift operation is carried out to generate a 64-bit linear array. Further operations like shift left, multiplication, AND, OR, and XOR results in blocks that are finally divided into 2^2 bits, followed by the vertical organization of the bits to ultimately generate the 4-bit keys.[6]

The next step is the key management protocol that uses a protocol—Localized Encryption and Authentication Protocol (LEAP) that employs a predetermined key to coordinate between the different varieties of keys that are associated with every node point.[6] Establishment of an individual node begins with a seed function where every node had a specific identification and to able to detect the respective node identification, a neighborhood detection technique procedure is employed. To compute the shared key amongst the adjacent nodes, the seed function along with an initial key is used by the receiver node that leads to the generation of any intermediate key whereby the redundant initial keys are removed. This allows for paired communication and is used to distribute the cluster key which is finally communicated by the base station. The use of the LEAP protocol guarantees a communication channel that is secure and protected and this step is followed by the encryption process.

Confusion and diffusion are created by performing logical operations, substitution, and swapping during encryption. The 64-bit array of input data is divided equally into two sections that are further subdivided into two more equal parts that are exchanged in subsequent rounds in order to generate a complex cipher. The output of each round acts as an input for the subsequent round which is eventually mapped into the *F*-function as expressed in the following equation[6]

$$F = \text{OR}(S - \text{boxes}(\text{AND}(\text{LS}(16\,\text{bits}\,/\,4)))) \quad (7.1)$$

The *F*-function prompts data dispersion with the use of logical AND/OR, shift, and *S*-boxes operations on the data. The *F*-function output is XORed and the exchanged data of the particular round result is used to create the data confusion that concludes the process of encryption.[6] The final step is the decryption process that is applied to retrieve the plaintext.

FIGURE 7.1 Reception of transmitted data (image) using SF algorithm.

The SF algorithm can be used in image data transmission where the image (input data) is converted to secret text and after decryption, the extracted image is obtained (as seen in Fig. 7.1).

7.3.1.4 ADVANCED ENCRYPTION STANDARD

This National Institute of Standards and Technology (NIST) algorithm has been used for IoT[11] and it comprises a fixed block with a maximum size of 256 bits and the size of the key starting from 128 bits to no theoretical maximum. The input is given by the user in a matrix format known as the state matrix on which multiple encryption rounds convert the input data to secret text. The design of AES (advanced encryption standard) algorithm consists of four steps as discussed subsequently.

7.3.1.4.1 SubBytes Step or Inv. SubByte Step

This step is used to introduce confusion of data in the input data matrix and it is the first step of iterative round transformation. The Inv. SubByte step is used for decryption process that is the inverse principle of SubByte step (as seen in Fig. 7.2).

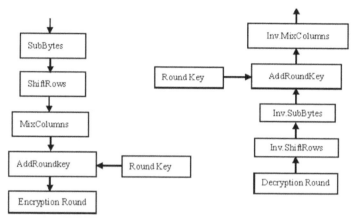

FIGURE 7.2 AES encryption and decryption process for one round.

7.3.1.4.2 ShiftRows Step or Inv. ShiftRows Step

Here, the state matrix bytes (each row) undergo cyclic shifts by a certain offset. While the bytes of the first row are not changed, each byte of the subsequent row undergoes one left shift (i.e., second row by one position, the third and fourth rows by two positions and three positions, respectively). Figure 7.6 shows the AES encryption and decryption process for one round. This step is used for decryption process that is the inverse principle of ShiftRow step.

7.3.1.4.3 MixColumns Step or Inv. MixColumn Step

In this step, each column (consisting of four bytes each) of the state matrix undergoes an invertible transformation. The polynomial that is generated in this linear transformation in a random fashion is arranged in a 4 × 4 matrix that is used during decryption process. XORing of each state matrix column with the equivalent column of the polynomial matrix is carried out and the result is updated in the same column and the final output matrix is fed as an input to AddRoundKey step. The Inv. MixColumn step is used for decryption process that is the inverse principle of MixColumn step.

7.3.1.4.4 AddRoundKey

The generation of the first round keys is done by an XOR operation between the initial (zeroth) round key and the state matrix and this is used for the generation of the next key for the subsequent round. The ciphertext is the output after multiple such rounds where Rijndael's key scheduling algorithm is used to generate the new round key for every round.

7.3.1.5 DATA ENCRYPTION STANDARD

This algorithm encrypts a 64-bit input data by making use of a 56-bit key to produce a ciphertext block of the same size as the input data. An effective key (56-bit) is generated from the 64-bit block for encryption–decryption. Owing to probable of brute-force attack, the once influential DES (data encryption standard) is now regarded to be insecure.

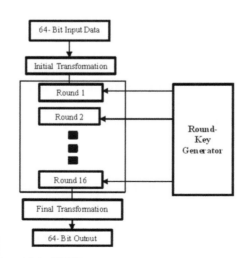

FIGURE 7.3 Basic model of DES.

Encryption and Decryption Algorithms for IoT Device Communication   105

The DES encryption and decryption have a difference in terms of the subkeys in a way that they are contrary to each other. The key size 64-bit can encrypt only 64-bit plaintext at a time (as seen in Fig. 7.3). Each 64-bit message can be encrypted using the same 64-bit key value using DES encryption operation. The diagram presents that 16 rounds are required to complete the whole process.

7.3.1.6   TRIPLE DATA ENCRYPTION STANDARD

This algorithm is similar to DES, which is applied thrice on the dataset to generate three subkeys each 64-bit long, from the user key. The steps involved in encryption are similar to the regular DES, but it is repeated thrice and hence, the name is Triple DES. While the first key finds its application in the encrypted process, the second key finds its use in the process of decryption, and finally, the third key is used for encryption again. The runtime of Triple DES is lesser than DES, but it offers higher security. In the Triple DES encryption process, the process starts with DES decryption, followed by DES encryption, and then DES decryption again.

7.3.1.7   BLOWFISH ALGORITHM

Another private-key algorithm that encrypts a data block of 64 bits at a time is the Blowfish algorithm that is an excellent choice in IoT for its lightweight and secure nature.[4] Blowfish uses a large number of subkeys that use the key expansion step to convert the 448 bits into different subkey arrays. The data encryption is a combination of permutation and substitution which are dependent on the key and the data, respectively. The operations are a combination of exclusive OR and addition on 32-bit words.

7.3.1.8   LIGHTWEIGHT ALGORITHMS

The development of better cryptographic algorithms to handle decreasing device size and RF (radio frequency) capability is an ongoing research trend and in order to handle these features, AES-based lightweight cryptography algorithms[12] are used as it is a NIST standard. An ultralightweight algorithm PRESENT[6,13] having a key size of 80–128 bits can run on a block of size 64 bits. It uses swapping and permutation operations and is hardware efficient. Owing to its time-consuming software cycles, an improvised version of this algorithm, RECTANGLE[14] was developed that is suitable for both minimized area designs and effective software implementations. It supports a block and key size of up to 64 and 128 bits, respectively. Another private-key block cipher CAMELLIA is designed for using in network layer as well as in hardware implementations like smart cards and it supports a 128-bit block size and working with up to keys length of 256 bits. A Feistel structure–based algorithm, namely, HIGHT[6,15] supports a 64-bit block size and CLEFIA[16] another Feistel structure–based algorithm that supports up 128-bit block size with up to 256-bit key lengths has also been tested for IoT. In comparison to other block ciphers, it provides superior security capability with an improved hardware performance. A notable Feistel structure–based algorithm for embedded networks is the TWINE.[17] SIMON[18] and SPECK[18] are other Feistel network structure–based algorithms that use basic operations like add, mod28, or logical XOR having a 64-bit block size, 128-bit key with 32 rounds. Other block ciphers[19] include Grasshopper, 3-Way, SHARK, SAFER, and Square that make use of swapping–permutation network that uses multiple sequences of swapping and transposition alternatively in order to ensure a change of secret text in a random manner. Secure IoT[20] is another private-key block cipher that has found in use in IoT.

7.4 WATERMARKING

Another method of adding security of device communication in the field of IoT is watermarking[21] where an image or text can introduce some information into the input data that may be invisible but may be understood with watermarking techniques. For data integrity protection, that is, in order to know if data has been tampered with or not, we use fragile watermarking. The data that can tolerate some change in the watermarked data caused by introduction of noise during compression makes use of semifragile watermarking. When recovery of the watermark is possible through suitable decoding techniques, it is called invisible–robust watermark and when the embedded watermark gets destroyed by any attacks, it is called invisible–fragile watermark. The watermarking procedure[5] includes the

Encryption and Decryption Algorithms for IoT Device Communication

process by which the watermark is inserted into the original data, and then it undergoes channel attacks, the watermark is extracted.

It begins with watermark embedding and when the data undergoes any kind of processing like compression, cropping, filtering, or any intentional attack like forgery, it may lead to attack on the watermark. The watermark embedding and attacks on the watermark can be seen in the following figures.

FIGURE 7.4 Example of a watermarked image.

How different attacks like speckle noise, blurring, or cropping operation affect the watermarked image (as seen in Fig. 7.4) can be seen in Figure 7.5 (a–c).

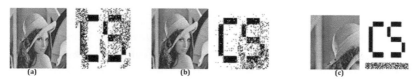

FIGURE 7.5 Different types of attacks used for watermark: (a) noise by speckle, (b) noise using blurring, and (c) noise by cropping.

The use of cryptography algorithms in order to recover an original data that has been affected by the process of blurring is possible. Here, we also see the original image and the effect of blurring on it and the extracted or recovered image after the watermark has been detected.

7.5 STEGANOGRAPHY ALGORITHMS

Steganography ("stegos" = "cover" and "grafia" = "writing") is yet another method that finds use in IoT security implementation where the secret data is hidden in a cover.[22] While in cryptography messages are scrambled so that unauthorized user fails to understand the message, in steganography, messages are hidden in a manner that its existence remains anonymous. Message and carrier are the fundamental features in steganography where the message represents the hidden data while data that the carrier carries may be text, image, or audio.

Steganography types[23,24] include text, image, and audio. Different types of coding schemes are used in text and audio steganography.[24,25] Image steganography uses the least significant bit operations on different types of image files.

In the text steganography,[26] a random character sequence is used to produce the cover text, jumbling the letter within a script or by altering the format of an existing data to hide the information. An image steganography[27] uses digital hiding methods to hide the image that can only be recovered on proper decoding. Audio steganography[25] uses noise or nonaudible frequency to hide the audio message.

7.6 HYBRID ALGORITHM IMPLEMENTATION

Data security and antidata corruption is a major concern and area of research in data communication applications in IoT device communication prior and post to data analysis. So there is a need for encryption–decryption process that goes under cryptography technique or watermark technique or steganography technique that is explored to achieve this. The data security in IoT device communication can be made by implementing a hybrid of watermarking and an encryption algorithm to get an enhanced security. Moreover, a feature for password may be included so that the authenticity of the receiver is verified before the data can be decrypted at the user end.

The whole process of hybrid technique using watermarking and cryptography algorithm implemented using Matrix Laboratory is presented in the next figure. Here, the hybrid technique is implemented by using watermark technique and cryptography technique (SF algorithm). Security 1 and 2 are the passwords in the form of one-time password or biometric

Encryption and Decryption Algorithms for IoT Device Communication 109

data and date of birth, respectively, that is specific to the authorized user. Different hybrid algorithm implementation using different watermarking techniques and cryptography algorithms have been reported in Ref. [28].

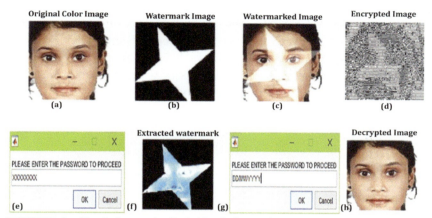

FIGURE 7.6 Simulation result of hybrid technique using watermarking and SF algorithm: (a) original image, (b) cover (watermark) image, (c) watermarked image, (d) encrypted image, (e) password box1, (f) extracted image, (g) password box2, and (h) decrypted image.

A hybrid technique using watermarking and SF algorithm is seen in Figure 7.6 where the input data is an image and a watermark is inserted into it. After this step, the watermarked image is encrypted. At the receiver end, the first password has to be given to extract the watermark while the second password is given to decrypt the final information. The implementation of a hybrid algorithm may prove to enhance the security in IoT device communication where we can inculcate the advantages of each algorithm to produce a robust method of securing communication between devices in an IoT network.

7.7 CONCLUSION

Security in communication over IoT networks can no longer be an afterthought. Some kind of security algorithm or techniques must be implemented in parallel with the implementation and upgradation of IoT networks. Security of the devices, security during the communication between these devices, and security of the data stored in the servers are all a part of the challenge that must be handled from the foundation when implementing an IoT-based network. This chapter gives a brief idea of different types of algorithms to maintain data security in IoT device communication. A hybrid algorithm was also presented, which lays the foundation of developing more such hybrid algorithms to make the communication between devices more secure. Some other solutions to handle the issues related to security in IoT can also be implemented by digital signature and digital fingerprints in which authentication and identification process of devices can be done.

KEYWORDS

- **IoT**
- **cryptography**
- **watermarking**
- **steganography**
- **encryption**
- **decryption**

REFERENCES

1. Stergiou, C.; Psannis, K. E.; Kim, B.-G.; Gupta, B. Secure Integration of IoT and Cloud Computing. *Future Gen. Comput. Syst*. **2018,** *78,* 964–975.
2. Nguyen, K. T.; Laurent, M.; Oualha, N. Survey on Secure Communication Protocols for the Internet of Things. *Ad Hoc Netw.* **2015,** *32,* 17–31.
3. Khan, M. A.; Salah, K. IoT Security: Review, Blockchain Solutions, and Open Challenges. *Future Gen. Comput. Syst*. **2017**.11.022; DOI: https://doi.org/10.1016/j.future.
4. Suresh, M.; Neema, M. Hardware Implementation of Blowfish Algorithm for the Secure Data Transmission in Internet of Things. *Global Colloquium Recent Adv. Effect. Res. Eng. Sci. Technol. (RAEREST 2016), Procedia Technol.* **2016,** *25,* 248–255.
5. Mishra, S.; Dastidar, A.; Samal, U. C.; Mohapatra, S. K. Review of Different Image Data Hiding Techniques. *Int. J. Adv. Sci. Technol.* **2018,** *121,* 43–54; http://dx.doi.org/10.14257/ijast.2018.121.04.

6. Mishra, S.; Dastidar, A. *Hybrid Image Encryption and Decryption Using Cryptography and Watermarking Technique for High Security Applications*. **2018**, 1–5; DOI: 10.1109/ICCTCT.2018.8551103.
7. Cirani, S.; Ferrari, G.; Veltri, L. Enforcing Security Mechanisms in the IP-Based Internet of Things: An Algorithmic Overview. *Algorithms* **2013**, *6*, 197–226.
8. Yu, H.; Kim, Y. New RSA Encryption Mechanism Using One-Time Encryption Keys and Unpredictable Bio-Signal for Wireless Communication Devices. *Electronics* **2020**, *9*, 246; DOI: 10.3390/electronics9020246.
9. Shantha, A.; Renita, J.; Edna, E. N. In *Analysis and Implementation of ECC Algorithm in Lightweight Device, International Conference on Communication and Signal Processing (ICCSP)*, Chennai, India, 2019; pp 0305–0309.
10. Vahdati, Z.; Ghasempour, A.; Salehi, M.; Yasin, S. Md. Comparison of ECC and RSA Algorithms in IoT Devices. *J. Theor. Appl. Info. Technol.* **2019**, *97* (16), 4293–4308.
11. Kumar, P.; Rana, S. B. Development of Modified AES Algorithm for Data Security. *Optik*. **2016**, *127*, 2341–2345.
12. Eisenbarth, T.; Kumar, S. A Survey of Lightweight-Cryptography Implementations. *IEEE Design Test Comput.* **2007**, *24* (6), 522–533.
13. Bogdanov, A. et al. PRESENT: An Ultra-Lightweight Block Cipher. In *Cryptographic Hardware and Embedded Systems – CHES*. Lecture Notes in Computer Science (Vol. 4727),Springer, 2007; pp 450–466.
14. Zhang, W.; Bao, Z.; Lin, D. et al. RECTANGLE: A Bit-Slice Lightweight Block Cipher Suitable for Multiple Platforms. *Sci. China Inf. Sci.* **2015**, *58*, 1–15.
15. Hong, D. et al. HIGHT: A New Block Cipher Suitable for Low-Resource Device. In *Cryptographic Hardware and Embedded Systems—CHES 2006*. Lecture Notes in Computer Science; 2006; pp 46–59.
16. Batra, I; Luhach, A. K. Analysis of Lightweight Cryptographic Solutions for Internet of Thing. *Indian J. Sci. Technol.* **2016**, *9* (28), 1–7.
17. Bhardwaj, I.; Kumar, A.; Bansal, M. In *A Review on Lightweight Cryptography Algorithms for Data Security and Authentication in IoTs, 2017 4th International Conference on Signal Processing, Computing and Control (ISPCC)*; Solan, 2017; pp 504–509.
18. Beaulieu, R. et al. In *The SIMON and SPECK Lightweight Block Ciphers*, Proceedings of the 52nd Annual Design Automation Conference, 2015; pp 1–6; DOI:10.1145/2744769.2747946.
19. Sadkhan, S. B.; Salman, A. O. In *A Survey on Lightweight-cryptography Status and Future Challenges, 2018 International Conference on Advance of Sustainable Engineering and Its Application (ICASEA)*; Wasit, **2018**; pp 105–108.
20. Usman, M.; Khan, S. SIT: A Lightweight Encryption Algorithm for Secure Internet of Things. *Int. J. Adv. Comput. Sci. App.* **2017**, *8*; DOI: 10.14569/IJACSA.2017.080151.
21. Kekre, H. B.; Athawale, A.; Rao, S.; Athawale, U. Information Hiding in Audio Signal. *Int. J. Comput. Appl.* **2010**, *7*; DOI: 10.5120/1278-1623.
22. Nosrati, M.; Karimi, R.; Hariri, M. An Introduction to Steganography Methods. *World Appl. Progr.* **2011**, *1*, 191–195.
23. Kumar, H. B. B. Digital Image Watermarking: An Overview. *Orient. J. Comput. Sci. Technol.* **2016**, *9* (1); **DOI** : http://dx.doi.org/10.13005/ojcst/901.02.

24. Chandran, S.; Bhattacharyya, K. In *Performance Analysis of LSB, DCT, and DWT for Digital Watermarking Application Using Steganography*, International Conference on Electrical Electronics Signals Communication and Optimization (EESCO), 2015; DOI: 10.1109/EESCO.2015.7253657.
25. Jian, C.; Chuah, C. W.; Rahman, N.; Hamid, A.; Isredza, R. In *Audio Steganography with Embedded Text*, IOP Conference Series: Materials Science and Engineering, 2017; p 226; DOI: 012084. 10.1088/1757-899X/226/1/012084.
26. Arora, H.; Bansal, C.; Dagar, S. *Comparative Study of Image Steganography Techniques*. 2018, 982-985. DOI: 10.1109/ICACCCN.2018.8748451.
27. Hussain, M.; Wahab, A. W. A.; Idris, I. B.; Ho, A. T. S; Jung, K.-H. Image Steganography in Spatial Domain: A Survey. *Signal Process.: Image Commun.* **2018**, *65*, 46–66.
28. Mishra, S.; Dastidar, A.; Nanda, I. Password Protected Image Encryption and Decryption Using Hybrid Technique for Data Security. *Curr. Trends Signal Process. (CTSP), STM J.* **2019**, *99* (1), 17–25.

CHAPTER 8

Future of Video Forensics in IoT

SUNPREET KAUR NANDA[1,2] and DEEPIKA GHAI[1*]

[1]*School of Electronics and Electrical Engineering, Lovely Professional University, Punjab, India*

[2]*EnTC Department, P. R. Pote College of Engineering and Management, Amravati, India*

*Corresponding author. E-mail: deepika.21507@lpu.co.in

ABSTRACT

Today is the hour of Internet of things (IoT), as countless machines, for instance, vehicles, smoke alarms, smart wristwatches, smart glasses, and webcams, are being related to the Internet. The number of machines that have the feature of remote access to screen and accumulate data is endlessly growing. This improvement makes, on the one hand, the human life progressively pleasant and supportive, although it moreover raises, on other hand, issues on security. Nevertheless, this progression furthermore raises troubles for the modernized operator when IoT devices are studied for criminal investigations. All the while, IoT computerized video legal devices are not reinforced by existing progressed quantifiable contraptions and methodologies, making it difficult for experts to extract data from them without the assistance of a legitimate advisor with explicit data at the present time. Without a doubt, current research asks about the composing bases on security and insurance for IoT circumstances, instead of methodologies or techniques of video forensics obtainment or examination for IoT devices. Different edges related to IoT forensics, scene examination through video recording, and variables of IoT influencing crime scene investigation are discussed in this chapter. In this chapter, we are also emphasizing on the IoT video forensic applications, limitations in the currently available video forensic tools, as well as the future of this research.

8.1 INTRODUCTION

At the point when a reader googles "IoT," a large number of the appropriate results are superfluously shown. A valid example is "The Internet of Things (IoT) is an arrangement of interrelated processing gadgets, mechanical and computerized machines, articles, creatures or individuals that are given one of a kind identifiers and the capacity to move information over a system without expecting human-to-human or human-to-PC association."[1]

At the present time, readers are perusing on their personal computers, cell phones, and possibly tablets, yet whatever gadget the users are utilizing, it is probably associated with the Web—the Internet. A Web affiliation gives all sorts of points of interest that essentially were not possible beforehand, and hence, it is an extraordinary thing.[2] In case you are adequately experienced, consider your mobile phone. When it was a basic phone, the user could only make calls and send text messages; however, at this point, the user can read any book, watch any film, or check out any music, all in their palm. The reality here is that the supporting things to the Internet yield several bewildering focal points. The users have watched the favorable circumstances with their advanced cell phones, workstations, and tablet phones; anyway this is legitimate for everything else.[3] Moreover, without a doubt, here it means the world. IoT is actually a very fundamental thought. It infers taking all the physical spots and all the things on the planet and interfacing them to Web. Confusion arises not because of the thought is so confined and immovably described, yet rather considering the way that it is so far-reaching and roughly portrayed. It might be hard to nail down the thought in someone's mind when there are such countless models and potential results in IoT. To help explain its imperative, let us comprehend the advantages of interfacing things to the Web.

8.1.1 IOT—ITS IMPORTANCE

When a device is connected with the Web, it can send or receive information, or both. This ability to send and receive information makes things "magnificent." An example of basic mobile phones again will be of great use. At this time, the user can look at basically any tune on the planet, yet it is not that their phones genuinely have each and every tune on the planet inside it. Considering the way that each tune on the planet is stored elsewhere, and in any case, their mobile phones can send data (streaming that tune) and sometime later get the data (downloading that tune on their mobile phones). To be smart, a thing may not have a super PC inside it—it fundamentally needs access to it.[4] In the IoTs, all the things that are being connected with the Web can be set into the following three requests:

(1) Things that aggregate data and send it sometime later.
(2) Things that get data and send them one after another sometimes later
(3) Things that do both.

Likewise, the entirety of the three of these has titanic ideal conditions that compound on one another. With the advancement in IoT and after connecting things to Internet, never again do we have to make sure to water the plants at home or farm, feed the pets, switch off lights and fans, or get back to check whether the car garage entryway was shut. Simultaneously, cutting edge innovations from our versatile and associated gadgets can play out a medical clinic grade electrocardiogram (ECG) from a faraway area and forward that information to the primary care physicians before we land at emergency clinic. Businesses have started to understand these new gadgets from both interruption and control. No dependable crime scene investigation application or advanced legal sciences direction exists to recover the information from IoT gadgets in case of a digital occasion, a functioning examination, or a case demand.[5] However, a colossal extending hole exists in our industry, organizing and executing the parts of the IoT. Each new gadget we make, each sensor we send, each byte we synchronize to different areas will sooner or later go under examination over the span of examinations and legitimate issues. The very rule that we should "secure" the gadgets infers that we will have the option to precisely decide whether the gadgets have been undermined. Not only exclusively does the direction not exist, but also the industry does not have a clue what information is caught in many occasions, what different gadgets the information arrives on, or if the information is meaningful and open or there is a chance that it would be recovered. The computerized video forensics of IoT innovations is the lost around discussion in our quick race to the guarantee of associating each gadget on this planet.[6] The remaining part of this chapter is organized as IoT forensics discussed in Section 8.2. Section 8.3 explains IoT video forensic applications. Section 8.4 discusses about related work in IoT forensics. We discuss novel variables of IoT influencing crime scene investigation in Section 8.5, whereas the need for video forensics in IoT is discussed in Section 8.6. Then, we enumerate and outline the IoT video forensics—challenges in Section 8.7, which is followed by a discussion about limitations in the currently available video forensic tools and future scope in Section 8.8. Finally, the conclusion is given in Section 8.9.

8.2 IOT FORENSICS

IoT legal sciences is a part of advanced crime scene investigation that manages IoT-related cybercrimes and incorporates examination of associated gadgets, sensors, and the information put away on every single imaginable stage. On the off chance that the user takes a gander at the master plan, IoT crime scene investigation is much progressively perplexing, multifaceted, and multidisciplinary in approach than customary criminology.[7] With adaptable IoT gadgets, there is no particular strategy for IoT crime scene investigation that can be comprehensively utilized. So distinguishing significant sources is a significant test. The whole examination will rely upon the idea of the associated or savvy gadget set up. For instance, proof could be gathered from fixed home computerization sensors, moving vehicle sensors, wearable gadgets, or information stored on cloud. Associated physical gadgets, home computerization machines, and wearable gadgets are all piece of IoT. This additionally carries with it a lot of chances for gigantic information ruptures and united digital security dangers.[8] The intention of computerized crime scene investigation is to recognize, gather, dissect, and present advanced proof gathered from different mediums in a cybercrime episode. The augmentation of IoT gadgets and the expanded number of digital security occurrences has brought forth IoT legal sciences. The process used in digital forensics[8] is shown in Figure 8.1.

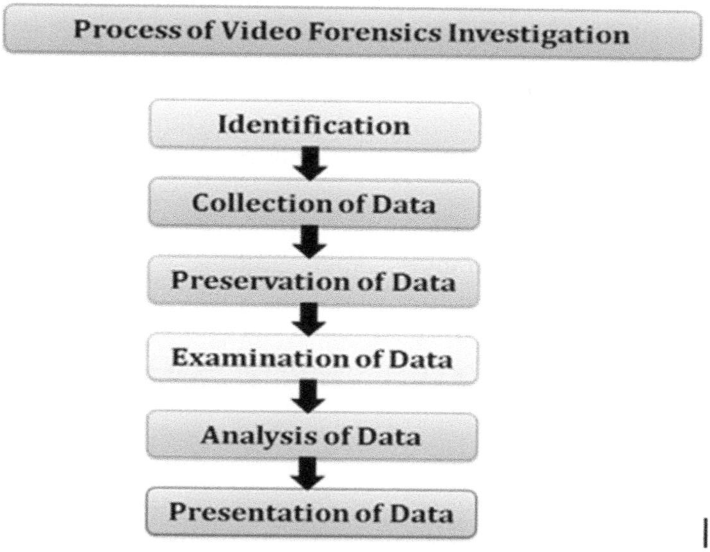

Process of Video Forensics Investigation

- Identification
- Collection of Data
- Preservation of Data
- Examination of Data
- Analysis of Data
- Presentation of Data

FIGURE 8.1 Process used in digital forensics.

The wellsprings of evidence in IoT crime scene investigation incorporate home machines, autos, medicinal inserts, sensor hubs, and label readers, along with others. In conventional criminology, the wellsprings of proof may be PCs, cell phones, servers, or passages. With respect to evidence of information, IoT information can be accessible in any merchant explicit organization, not at all like in conventional crime scene investigation wherein information is generally accessible in an electronic record or standard document positions.[9] IoT crime scene investigation examination includes numerous difficulties due to the nonreasonableness of present accessible computerized criminology instruments and standard legal sciences techniques in the IoT condition. IoT gadgets additionally create a gigantic measure of different information that bewilders examiners when choosing the pertinent wellspring of proof and distinguishing the specific measure of information to be utilized for additional examination. A few studies were led beforehand on the utilization of computerized criminology in various spaces, that is, distributed computing, edge figuring, portable distributed computing, programming characterized systems, remote systems, smart urban areas, and savvy transportation frameworks. Notwithstanding, none of these studies exhaustively centered on IoT legal sciences. Likewise, a few other significant parts of IoT crime scene investigation, which are talked about in the present examination, have not been recently revealed.[10]

8.3 IOT VIDEO FORENSIC APPLICATIONS

Treatment of case identified with the utilization of data innovation regularly requires crime scene investigation. Forensic is an action to direct examinations and set up realities identifying with criminal occasions and other legitimate issues.[11] Advanced crime scene investigation is a piece of measurable science, enveloping the disclosure, and examination of the material (information) could be seen in Figure 8.2.

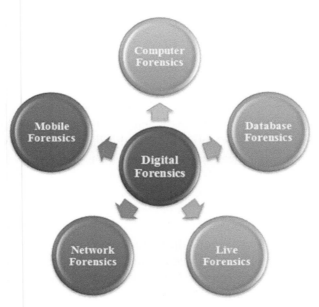

FIGURE 8.2 Uses of advanced digital forensics.

For most, "forensics or criminology" infers great procedural shows like bones, criminal minds, or cold case. Possibly the user pictures white covers in a lab, tidying normal family objects for fingerprints. Measurable science is the use of science to both crook and common law. Such exertion reveals both unmistakable and fair proof in deciding the result of legitimate procedures.

With the advancement of the Web, it is seen that the development of criminal conduct and computerized wrongdoing requires an interesting and novel type of legal sciences skill. In the event that somebody exploits their conduct online to take the cash, or more awful, their personality, specialists utilize computerized crime scene investigation to clean for advanced fingerprints to get the troublemaker.[12–15] The data stored on a storage device is mostly confidential or private, so the data must be secured from a third party.

8.3.1 INTELLIGENT TRANSPORT SYSTEM

Singapore is utilizing smart transportation frameworks shown in Figure 8.3. Presently, smart sensors and different gadgets are sent to oversee traffic, what is more, keep away from traffic blockage issue. Precision is one of the most critical parameters that must be considered in adroit transportation structure. Insufficient and wrong information can cause great disasters in the city.[16] On account of mishap, the legal sciences specialist is required to realize what and how something went wrong. The examination can help alleviate mishap causing issues or on the other hand different issues, for example, traffic blockage. This can in turn be looked upon by installing CCTV cameras at traffic signals.

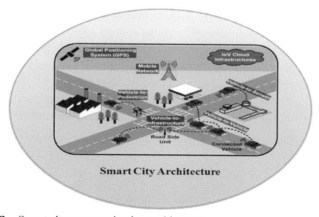

FIGURE 8.3 Smart city communication architecture.

8.3.2 TYPICAL SMART HOME

IoT, however, presents another degree of digital risk.[17-22] Your web conduct never again exists inside the bounds of your work area or cell phones; it presently incorporates your smart home,[22] associated vehicles, and numerous different gadgets such as light, fan, and computer, as shown in Figure 8.4.

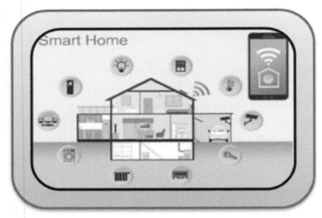

FIGURE 8.4 Modern smart home.

Indeed, even your smart pet collar opens the entryway for unauthorized people to get to your information. Even the camera installed at different locations in the home will have all the data in the form of video for investigation, if required. IoT forensics endeavors to adjust its work process to that of any legal sciences practice—agents distinguish, decipher, protect, examine, and present any significant information. Like any examination, a course of events is developed, and, with the guidance of savvy gadgets' information, agents may have the option to catch substantially more explicit information focuses than in a conventional wrongdoing.[13]

8.3.3 ECHO GADGET

Specialists should likewise consider the job of IoT inside the setting of the wrongdoing. The IoT gadget is not the only apparatus to carry out the wrongdoing or an observer to the wrongdoing. In a 2015 crime in Bentonville, Arkansas, police gathered an Amazon Echo gadget from the home where the homicide occurred (Fig. 8.5).[23] Since Echo gadgets can in some cases get and record incomplete discussions (just as express directions), the police gave a warrant to Amazon for any recorded information from the gadget.

FIGURE 8.5 Echo gadget handed over by Amazon for crime investigation.

Correspondingly, in a 2017, twofold crime in Farmington, New Hampshire, a judge gave a warrant requesting Amazon to discharge 2 days' worth of information from an Amazon Echo, in trusts that it recorded piece of the assault. In the two occasions, Amazon was eminently safe as well as delayed to react in the discharging of client information to specialists.[14]

8.4 RELATED WORK

Various advanced scientific models have been delivered for assessments; a segment of these focused on either scene, response or assess or stress a particular stage or development of an assessment. Comparative summary of some possible IoT forensic systems are shown in Table 8.1.

8.5 NOVEL VARIABLES OF IOT INFLUENCING CRIME SCENE INVESTIGATION

Various new factors of IoT influencing conventional PC crime scene investigation[10] are sketched out in Figure 8.6. Countless differing and asset compelled gadgets are engaged with IoT-empowered conditions, which produce a tremendous measure of information known as "Big IoT Data." Lots of IoT information forestall the criminology examiner to gather and concentrate the proof information easily. The primary difficulties presented by big IoT data that too for the legal science agents are assorted information arrangements and absence of ongoing log examination arrangements. Advanced proof is one of the central necessities for empowering IoT crime scene investigation. Such a proof must be gotten by separating firmware information or securing a blaze memory picture. As far as computerized proof is concerned, restricted perceivability and short retention time of the confirmations are the new difficulties presented by the IoT gadgets that influence the conventional PC scientific answers for being applied in the IoT frameworks. In the typical situations, information are for the most part put away and prepared on the cloud. Much of the time, getting access to information for examination purposes gets hard for IoT crime scene investigation specialists because of administration-level understanding requirements.

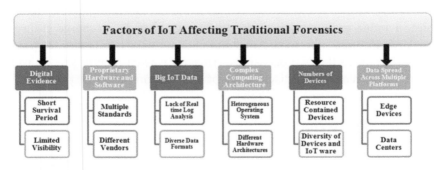

FIGURE 8.6 Novel components of IoT influencing conventional PC forensics.

TABLE 8.1 Comparative Summary of Some Possible IoT Forensics Systems.

Author (Year)	Use	Detail	Possible proof	Application of IoT	Video forensics needed	Region
Oriwoh et al. (2013) [1]	Human internal organs and home appliances will be enabled to communicate	For checking presence of medical implants in plants as well as animals	Identifying the next best possible thing as an evidence	Medical implantations to be done on humans and animals	Yes	Global
Perumal et al. (2015) [3]	Radical digitalization of industries	Connecting the new technology with sources, people, and machinery	CCTV	Smart industry	Yes	Netherlands
Yaqoob et al., 2018 [9]	IoT-based vehicle system	Collecting and preserving highly important evidences	Smart vehicles	Internet of vehicles with trust IoV	No	Global
Servida et al. (2019) [12]	Network traffic investigation and traces on physical device	Increasing amount of network traffic being encrypted, traces of configuration settings	Mobile phone	iSmartAlarm CubeOne	Yes	Global
Javed et al. (2016) [15]	Detection of malware	Detection of malware in smart traffic environment	Cameras, traffic light system	Intelligent traffic system	Yes	Global

TABLE 8.1 (Continued)

Author (Year)	Use	Detail	Possible proof	Application of IoT	Video forensics needed	Region
Vippalapalli et al. (2016) [16]	Smart health monitoring system	Checking status of ones well-being using smart wearable devices	Smart wearable device such as smartwatch	Smart health caring systems	Yes	Global
Malche et al. (2017) [17]	In typical smart home for forensic edge management system	Measuring the edge or fringe level security	Smart home appliances like smart phone, smart TV, lighting system, and parking system	Smart home systems	Yes	Global
Sherly et al. (2015) [18]	Intelligent transportation system	Managing the routes in a smart and efficient manner	Smart vehicles, cameras, and satellite	Smart transportation system	Yes	Singapore
Tziortzioti et al. (2019) [19]	Flood defense system	Getting information about the water level so that flood can be avoided	Smart sea sensors	Smart sea monitoring system	Yes	United Kingdom

Moreover, the information of IoT condition is spread over various stages, for example, on the fringe gadgets and server farms. The calculation is done basically at the edge or fringe of clients' systems, and metadata is moved to the cloud. The information are put away in two pecking orders, which make challenges for legal science examiners regarding information assortment and log information examination in such situations. The two other IoT factors influencing video forensics are intricate figuring engineering (i.e., diverse equipment designs and heterogeneous working frameworks) and restrictive equipment and programming.

The IoT facilitates various sensors, articles, and sharp center points that are fit for talking with each other without human intervention. The things work autonomously with regard to various articles. IoT center points are fit for passing on lightweight data, finding a workable pace cloud-based resources for get-together and isolating data and choosing decisions by looking at assembled data. The ascent of IoT has incited unpreventable relationship of people, organizations, sensors, and things. IoT contraptions are as of now passed on in a wide extent of uses from savvy systems to therapeutic administrations and information transport structures. Tremendous business openings that exist inside IoT space in a general sense extended a number of sharp contraptions and sagacious, free organizations offered in IoT frameworks. Furthermore, reliance of IoT devices on cloud establishment for data movement, amassing, and assessment initiated improvement of cloud-engaged IoT frameworks. Security issues, for instance, assurance, find a good pace; correspondence and secure storing of data are ending up being important troubles in IoT condition. On the other hand, every single contraption that the user makes, each new sensor that is sent, and every single byte that is synchronized inside an IoT circumstance may sometime go under assessment. The snappy advancement of IoT contraptions and organizations initiated sending of various powerless and questionable center points. Additionally, standard customer-driven security structures are of little use in object-driven IoT masterminds. Therefore, explicit instruments, methodologies, and approach for checking IoT frameworks and assembling, ensuring, and researching remaining affirmations of IoT conditions are required.[7]

8.6 NEED FOR VIDEO FORENSICS IN IOT

It is anything but an extraordinary jump to envision wearable advances being utilized as verifying proof that an individual might be sleeping or practicing at the exact time of an occasion. It will be very disturbing when somebody's IoT home mechanization framework is crippled by some suspect to pick up section into a home or install sensors in the new IoT-prepared urban areas, catch much extra information, focus at the exact scene of a wrongdoing. It is difficult to come to know when an edge hub IoT gadget is undermined over the unprecedented system convention, increasing a solid footing into a current system. In these cases, computerized legal science experts will be called to recover information that may exist somewhere on these gadgets. The users look forward to a universe of extending omnipresent processing, and the tests are taking place at present. Readers cannot envision the actual numbers of end hubs, as they are growing at rates much higher than we have experienced to date. On the off chance that these gadgets are progressively defenseless on systems in view of juvenile security capacities, we can be guaranteed that examinations will be expected to comprehend what job these gadgets played in a breach.[6]

8.7 IOT VIDEO FORENSICS CHALLENGES

The exceptional augmentation of IoT contraptions, for instance, mobile phones, garments washers, and therapeutic aids, has enabled individuals to grant data to each other. These contraptions can chat with each other directly or through application programming interface over the Internet, and they can be constrained by "learned" gadgets with high dealing with limits, for example, cloud servers, that extension information to low-enrolling gadgets. The speed and correspondence limits of IoT contraptions offer different strong applications to basic occupants, affiliations, industry, and governments.[4] IoT application is in like manner loosened up during the zones of transportation, social protection, and sharp urban zones. Additionally, the market example of IoT is growing, as showed by CISCO's estimation of IoT salary, which will be around $14.4 trillion some place in the scope of 2013 and 2022. In any case, creating IoT developments faces diverse security attacks and risks. Striking threats consolidate

Future of Video Forensics in IoT

contamination attacks, mass surveillance, and denial of service ambushes, and unsettling influence of IoT frameworks. To investigate these attacks, solid and steady gatherings must lead an advanced assessment, known as IoT legitimate sciences, on the bad behavior scene.[10] A portrayal of security stresses in IoT-based sagacious circumstances[10] is given in Figure 8.7.

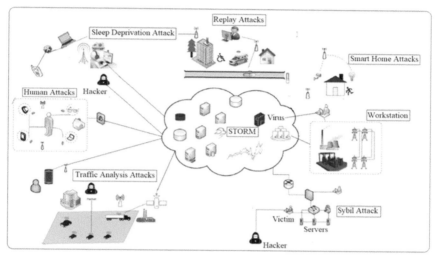

FIGURE 8.7 A delineation of security threats in IoT-based typical situations.

Numerous examinations are done on security of IoT, the writing on IoT legal sciences, or forensics is rare. Figure 8.8 displays the titles of distributed chips at IoT legal sciences.[10]

FIGURE 8.8 Word recurrence events in many distributions.

When contrasted with the standard computerized scientific systems, IoT forensics depicts numerous difficulties relying upon the flexibility and multifaceted nature of the IoT gadgets. Following are a few difficulties that one may look in an examination:

- fluctuation of the IoT gadgets,
- exclusive hardware and software,
- information present over numerous gadgets and stages,
- information can be refreshed, changed, or lost, and
- restrictive wards for information is put away on cloud or an alternate topography.

In that capacity, a multifaceted methodology is required by IoT, where proof can be gathered from different sources. We can arrange wellsprings of proof into three general gatherings[7]:

- Shrewd gadgets and sensors: gadgets present at the wrongdoing scene (smartwatch, home mechanization apparatuses, climate control gadgets, and that is the tip of iceberg)
- Equipment and software: the correspondence interface between smart gadgets and the outer world (PCs, portable gadgets, IPS, and firewalls)
- Outer assets: territories outside the system under examination (cloud, informal communities, ISPs, and versatile system suppliers).

When the proof is effectively gathered from an IoT gadget regardless of the document framework, working framework, or the stage it depends on, it ought to be logged and observed. The primary purpose for this is IoT gadgets' information stockpiling that is significantly on cloud because of its adaptability and availability. There are many potential outcomes, and the information on cloud can be modified, which would result to an examination disappointment. Presumably cloud crime scene investigation can similarly assume a significant job here, yet reinforcing digital security.

With regularly advancing IoT gadgets, there will consistently be a requirement for exceptional practice strategies and systems to get through the examination. Cybercrime continues advancing and getting bolder constantly. Criminology specialists should create ranges of abilities to manage the assortment and multifaceted nature of IoT gadgets to stay aware of this advancement. Regardless of the difficulties one faces, there is constantly an extraordinary answer for complex issues. There will consistently be a requirement for one of a kind, canny, and versatile methods to research IoT-related violations and a considerably more prominent requirement for those showing these capacities.

8.8 LIMITATIONS IN THE CURRENTLY AVAILABLE VIDEO FORENSIC TOOLS

The current apparatuses in advanced crime scene investigation field cannot match with the heterogeneous framework of IoT condition. The enormous measure of conceivable proofs that are created by countless IoT gadgets, it will thus get new difficulties the part of gathering proof from disseminated IoT foundations. Moreover, since a programmer can find out the proof in IoT gadgets, in light of the fact that there are shortcomings of these gadgets in term of security, the extricating proof from them are possibly not satisfactory in the court of law. In addition, in light of the fact that a large portion of IoT information are put away in the cloud, the cloud becomes one of the principle sources of proof in IoT.[11] Consequently, examiners will confront a portion of the issues of gathering proof from the cloud on the grounds that the methods of computerized criminological and devices accept to have physical access to the proof source. Be that as it may, in the cloud, the specialists could discover a trouble to even to know where the information is found. What is more, the physical servers could have numerous virtual machines that have a place with various proprietors. In addition, cloud conditions could not be accessed when a wrongdoing has been detected. Thus, these initiations should tend to discover a strategy to work around boundaries and think of another instrument for IoT examination, which can be affirmed by the court of law and accomplish of agents' objectives.[10]

8.8.1 FUTURE SCOPE

IoT innovation has displayed a critical move in examination field, particularly by the way it cooperated with information. Many of IoT information are spread in various areas that are out of the clients' control. This information could be in the cloud, in outsider's area, in cell phone, or different gadgets. Along these lines, in IoT crime scene investigation, distinguishing the area of proof is considered as perhaps the greatest test can a specialist perform so as to gather the proof.[8] What is more, IoT information may be situated in various nations and be blended in with different clients data, which implies unique nations' guidelines are included. An incredible case model is the thing that occurred in August 2014, when a Microsoft did not follow a court order that looked for information put away outside the nation of warrant United States), putting forth the defense opened for a significant stretch of time. Life expectancy restriction of computerized media because the constraint of storage in IoT gadgets, the life expectancy of information in IoT gadgets is short and information can be effectively overwritten, bringing about the plausibility of proof being lost. Along these lines, one of the difficulties is the time of retention of the proof in IoT gadgets before it is overwritten. Moving the information to something else, for example, nearby hub or to the cloud, could be a simple arrangement to settle this test. Notwithstanding, it presents another test that identified with tying down the chain of proofs, and how to demonstrate the proof has not been changed or altered. Most of the records of cloud administration are of anonymous clients since cloud administration does not require the exact data from client to pursue their administration. It could prompt difficulty in recognizing a criminal. For instance, despite the fact that the examiners discover a proof in the cloud that demonstrates a specific IoT gadget in wrongdoing scene is the reason for the wrongdoing, it does not mean this proof could help in recognizing the lawbreaker. Evidence in IoT gadgets could be changed or erased due to absence of security, which could make these proof not strong enough to be acknowledged in the court of law. For instance, in the market, a few organizations do not refresh their gadgets normally or at all, or at some point, they quit supporting the gadget's system when they center on another item with the new framework. Thus, it could leave these gadgets powerless as programmer found powerlessness. In recognizable proof period of crime scene investigation, the computerized agent requires to distinguish and get the proof from a computerized wrongdoing scene.[2,3] For the most part, proof source is a kind of a PC framework, for example, PC and cell phone. In any case, in IoT, the wellspring of proof could be objects like a smart fridge or espresso coffee maker. Subsequently, the specialists will confront a few difficulties. One of these difficulties is identifying and finding the IoT gadgets in wrongdoing scene. Conveying the gadget to the lab furthermore, finding a space could be another challenge that specialists could confront as far as gadget type. Likewise, separating the confirmation structure, these gadgets are considered as anther IoT challenges, as a large portion of maker embraces various stages, working frameworks, and equipment types.[1,5] One of the models is the CCTV crime scene investigation where the makers of CCTV applied distinctive record framework group in their gadgets. Recovering appropriately, relics from CCTV's stockpiling gadgets are yet a difficulty. Also, data could be processed, utilizing diagnostic works in better places before being put away in the cloud. Subsequently, so as to be acknowledged in a court of law, information structure ought to come back to unique arrangement before performing analysis.

8.9 CONCLUSION

While we realize that IoT gadgets gather, store, and offer important data to assist us with settling on close to home and business choices, even in these beginning times of associated gadgets we can see regularly advancing employments. Hence, from the comparative study shown in Table 8.1, we conclude that IoT is becoming important whenever video forensics is concerned. IoT when combined with video forensics can identify minute details of crime in a jiffy. The potential for these resources such as CCTV camera and mobile camera to empower new wrongdoings makes an altogether new sort of watchfulness for the general population, but then their capacity to record and spare data could demonstrate significantly for examiners with no different leads.[13] Driving innovation organizations likewise need to accommodate how they decide to help (or not help) specialists when their own gadgets might open answer in conceivably savage criminal cases. Despite the fact that coordinating different solid highlights in a product will make the examination procedure simpler and progressively exact, still there would be a requirement for additional exploration to propose upgraded video altering systems to accelerate the video altering process which will have grind sway on improving video forensics instruments.

KEYWORDS

- Internet of things (IoT)
- scene examination
- IoT forensics
- video forensics

REFERENCES

1. Oriwoh, E.; Jazani, D.; Epiphaniou, G.; Sant, P. In *Internet of Things Forensics: Challenges and Approaches*, 9th IEEE International Conference on Collaborative Computing: Networking, Applications and Worksharing; Austin, TX, USA, Oct 20–23, **2013**; pp 608–615; doi: 10.4108/icst.collaboratecom.2013.254159
2. Hegarty, R. C.; Lamb, D. J.; Attwood, A. In *Digital Evidence Challenges in the Internet of Things*, Proceedings of the 9th International Workshop on Digital Forensics and Incident Analysis; 2014; pp 163–172.
3. Perumal, S.; Norwawi, N. M.; Raman, V. In *Internet of Things (IoT) Digital Forensic Investigation Model: Top-Down Forensic Approach Methodology*, 5th International Conference on Digital Information Processing and Communications; Sierre, Switzerland, Oct 7–9, 2015; pp 19–23; doi: 10.1109/ICDIPC.2015.7323000
4. Liu, J. In *IoT Forensics Issues, Strategies and Challenges*, 12th IDF Annual Conference; Liu, Tokyo, Japan, Dec 15, 2015.
5. Lillis, D.; Becker, B. A.; Sullivan, T. O.; Scanlon, M. Current Challenges and Future Research Areas for Digital Forensic Investigation. *Cryptography Secur.* **2016**, 1–11; doi: 10.13140/RG.2.2.34898.76489
6. Kebande, V. R.; Ray, I. In *A Generic Digital Forensic Investigation Framework for Internet of Things (IoT)*, 4th International Conference on Future Internet of Things and Cloud; Vienna, Austria, Aug 22–24, **2016**;,pp 356–362; doi 10.1109/FiCloud.2016.57
7. Watson, S.; Dehghantanha, A. Digital Forensics: The Missing Piece of the Internet of Things Promise. *Comput. Fraud Secur.* **2016**, *6*, 5–8; doi: 10.1016/S1361-3723(15)30045-2
8. Conti, M.; Dehghantanha, A.; Franke, K.; Watson, S. Internet of Things Security and Forensics: Challenges and Opportunities. *Future Gen. Comput. Syst., Elsevier* **2018**, *78*, 544–546; http://dx.doi.org/10.1016/j.future.2017.07.060
9. Yaqoob, I.; Hashem, I. A. T.; Ahmed, A.; Kazmi, S. M. A.; Hong, C. S. Internet of Things Forensics: Recent Advances, Taxonomy, Requirements, and Open Challenges. *Future Gen. Comput. Syst., Elsevier* **2019**, *92*, 265–275; doi: 10.1016/j.future.2018.09.058
10. Justicia, A. P.; Riadi, I. Analysis of Forensic Video in Storage Data using Tampering Method. *Int. J. Cyber-Secur. Digital Forensics* **2018**, *7* (3), 328–335; doi: 10.17781/P002471

11. Alabdulsalam, S.; Schaefer, K.; Kechadi, T.; Khac, N. A. L. In *Internet of Things Forensics—Challenges and a Case Study*, 14th IFIP International Conference on Digital Forensic (Digital Forensics); New Delhi, India, **2018**, *532*, 35–48; doi: 10.1007/978-3-319-99277-8_3
12. Servida, F.; Casey, E. IoT Forensic Challenges and Opportunities for Digital Traces. *Digital Invest., Elsevier* **2019**, *28*, 522–529; https://doi.org/10.1016/j.diin.2019.01.012
13. Shahraki, A. S.; Sayyadi, H.; Amri, M. H.; Nikmaram, M. Survey: Video Forensic Tools. *J. Theor. Appl. Inf. Technol.* **2013**, *47*, 98–107.
14. Kyei, K.; Zavarsky, P.; Lindskog, D.; Ruhl, R. In *A Review and Comparative Study of Digital Forensic Investigation Models*, International Conference on Digital Forensics & Cyber Crime; 2012; pp 314–327; https://doi.org/10.1007/978-3-642-39891-9_20
15. Javed, M. A.; Hamida, E. B.; Znaidi, W. Security in Intelligent Transport Systems for Smart Cities: From Theory to Practice. *MDPI J. Sensors* **2016**, *16* (6), 879; doi:10.3390/s16060879
16. Vippalapalli, V.; Ananthula, S. In *Internet of Things (IoT) based Smart Health Care System*, IEEE International Conference on Signal Processing, Communication, Power and Embedded System; Paralakhemundi, India, Oct 3–5, 2016; pp 1229–1233; doi: 978-1-5090-4620-1
17. Malche, T.; Maheshwary, P. In *Internet of Things (IoT) for Building Smart Home System*, International conference on IoT in Social, Mobile, Analytics and Cloud; Palladam, India, Feb 10–11, 2017; pp 65–70; doi: 10.1109/I-SMAC.2017.8058258
18. Sherly, J.; Somasundareswari, D. Internet of Things Based Smart Transportation Systems. *Int. Res. J. Eng. Technol.* **2015**, *2* (7), 1207–1210.
19. Tziortzioti, C.; Amaxilatis, D.; Mavrommati, I.; Chatzigiannakis, I. IoT Sensors in Sea Water Environment: Ahoy! Experiences from a Short Summer Trial. *Electron. Notes Theor. Comput. Sci.; Elsevier* **2019**, *343* (4), 117–130.
20. Boricha, V. IoT Forensics: Security in an Always Connected World Where Things Talk; https://hub.packtpub.com/iot-forensics-security-connected-world/.
21. Dunn, R. IoT Applications in Forensics; https://www.iotforall.com/iot-applications-forensics/.
22. How to Design a Smart Home System; https://www.smartsecurity.guide/how-to-design-a-smart-home-automation-system/
23. After Pushing Back, Amazon Hands Over Echo Data in Arkansas Murder Case; https://techcrunch.com/2017/03/07/amazon-echo-murde

CHAPTER 9

Role of Microstrip Patch Antenna for Embedded IoT Applications

AMANDEEP KAUR*, PRAVEEN KUMAR MALIK, and RAVI SHANKAR

Department of Electronics and Communication Engineering, Lovely Professional University, Jalandhar, Punjab, India

*Corresponding author. E-mail: aman.dhaliwal18@gmail.com

ABSTRACT

Nowadays, the Internet of things (IoT) is the paradigm that is used to describe what is expected from industry in terms of connectivity of huge devices with each other. In this structure, sensors, actuators, microcontroller, etc. are connected with each other through the Internet and can share the data between them at anytime and anywhere. IoT initially was derived from radio-frequency technology in MIT (Massachusetts Institute of Technology). This field is seeking attention in every application domains like healthcare, transportation, agriculture, communication, home automation, space monitoring, etc. IoT network and interconnected devices use licensed and unlicensed frequency bands like 2.4-GHz ISM (Industrial Scientific and Medical) band with bandwidths less than 5 MHz. All devices in network are connected through star, mesh, or point-to-point technologies to exchange data through uplink and downlink. To transfer signals wirelessly, omnidirectional or directional antennas are needed like microstrip patch, wire, whip, rubber antennas used inside Bluetooth, Arduino boards with ESP8266, GPS, Wi-Fi modules. So, RF (Radio Frequency) signals travel based on some propagation characteristics that describe their quality. To exchange data through RF signals, antenna is used to transmit and receive signal on transmitter and receiver sides, respectively. Antenna plays predominant role in the growth of IoT market. To make embedded circuits more compact, smaller sized and lightweight antennas are desired. This tremendous growth in IoT field had placed demand over multiband and ultrawideband antennas. Proper antenna selection and integration with device affect the overall system performance in transmitting data at high rates while consuming less power. In this chapter, the role of microstrip patch antenna is discussed in embedded IoT application.

9.1 INTRODUCTION

This chapter explains the need for microstrip patch antenna in embedded Internet of things (IoT) applications. The IoT is the network of embedded devices like processors, analog or digital sensors, actuators, and a set of instructions called programs used to design hardware circuits and develop printed circuit boards; further to connect to the Web, they need APIs and protocols of embedded devices. To connect these devices wirelessly in a network to share data at high speeds, need efficient antennas to couple electrical signals without interferences. Moreover, to design efficient antenna systems, some key points are considered to play quintessential role like size, cost, speed, and power and consumption.

9.2 LITERATURE REVIEW

The literature describes some antennas used for IoT applications in detail with different specifications and applications with proper band of selection. In Ref. [1], microstrip patch antenna was designed for IoT applications. Conventional patch concept was used with two more slotted patches connected with main patch. Antenna was designed using Rogers RT/Duroid 5880 substrate material with dielectric constant 2.2 and thickness 3.2 mm having loss tangent 0.0009. Antenna structure was fed using coaxial feed. From simulation results, it was depicted that the antenna operates on two different resonate bands 2.39 and 3.15 GHz used for IoT applications. The author further analyzed its performance with/without making connections of side patches with main patch and found out gain variations from 4.4 to 5.01 dB as given in Fig. 9.1.

Role of Microstrip Patch Antenna for Embedded IoT Applications 137

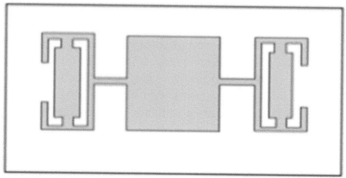

FIGURE 9.1 Slotted antenna design.

Frequency band used for RFID (Radio Frequency Identification) varies in different countries ranging from 850 to 950 MHz. For efficient communication between RFID reader and tag, antenna should have linear polarization. Various antennas are used for this purpose but all are bulky, so in Ref. [2], Vikram designed microstrip antenna with circular patch for RFID reader to operate in ISM band. Antenna was fabricated using RT/Duroid 5870 with εr = 2.33, loss tangent 0.001, and thickness t = 1.17 mm. Ground plan of proposed antenna was circular in shape with r = 23.1 mm and radius of circle on patch was 15.3 mm. Antenna structure was fed using microstrip line feeding technique. Results were simulated using HFSS (High Frequency Simulation Software) and it was analyzed that antenna efficiency is good, providing gain of 6.5 dBi with return loss of −40 dB and giving circular polarization.

In IoT applications, all devices are connected with each other, so IoT devices must have multiple connectivities. Wireless standards like LoRa and Sigfox transmit the data on low data rates and low-power networks fall under wide area network in terms of range but applications like GPS and Bluetooth are used for personal communication like mobile phones. So, multiple-band antenna is required to communicate over different frequency bands. This behavior can be achieved using devices like digital tunable capacitors, PIN diodes, and varactor diodes. Using this concept, in Ref. [3], Asadallah et al designed reconfigurable circular patch antenna with radius of 29.8 mm on Rogers RO5880 dielectric substrate with εr = 2.2 and thickness of 0.787 mm. From simulation results, it was analyzed that proposed antenna operates on three different bands, that is, 2.4 GHz,

1.58 GHz, and 868 MHz. Antenna resonates at LoRa, GPS, and Bluetooth bands with 50% reduced size.

In Ref. [4], Vyshnavi Das designed starfish-shaped antenna that resonates on three different frequency bands. Antenna fabrication was done on FR4 material with dielectric constant 4.4 and loss tangent 0.002. Microstrip feed line was used to excite antenna with patch dimension 110 mm × 90 mm with height 6.35 mm. Antenna simulation was carried out using HFSS software and it was analyzed that antenna operates on Bluetooth, Wi-Fi, and WiMAX bands with impedance bandwidth of 340, 120, 180, 300, and 240 MHz, respectively. It includes both IoT and cellular applications.

The author in Ref. [5] designed slotted patch microstrip patch antenna that operates in wireless local area network (WLAN). Rogers RT5880 substrate was used with dielectric value 2.2 and thickness 1.6 mm. Patch dimensions were 41.32 mm by 49.38 mm and simulated using CST software. Cross slotted patch was used for antenna bandwidth enhancement. From simulation results, it was depicted that antenna works on 802.11 b/n/g Wi-Fi standards for IoT applications.

In Ref. [6], circular patch antenna was proposed with dimensions of 30 mm by 20 mm. Antenna designing was done suing Rogers Ro4003 substrate with εr = 3.55 and loss tangent 0.0027. Antenna simulation results showed that it works for IoT and satellite applications like 1.8, 5.2, 7, and 8.53 GHz for RFID, WLAN, and C band, respectively, with bandwidth of 200 and 580 MHz.

For medical application in the field of IoT, small wearable antennas are needed. To improve antenna efficiency, metamaterial and fractal concepts are used. In Ref. [7], Sabban used this concept to design highly efficient antenna for medical and wearable applications. Antenna was designed on PCB (Printed Circuit Board) with dielectric constant 2.2 and patch dimensions 36 mm × 20 mm with thickness 3.2 mm. From simulation results, we came know that proposed antenna with split-ring resonator (SRR) has high gain and directivity as compared with antenna without SRR.

9.3 WHAT IS SYSTEM?

A system can be defined as a combination of subunits or subparts, integrated together to perform some work in well-organized manner according to a fixed plan, a set of instructions or programs. In system, all units are assembled and integrated well with each other to accomplish the desired goal. According to resulted output, system can have various inputs, which have to go through certain rules or processes to generate outputs. Interestingly, system consists of many subsystems, for instance, watch, washing machine, microwave oven, computer, and car, to proceed toward the overall desired result or goal.

Systems can be simple or complex in nature. Any fault in some parts of subsystem can lead to false outputs.

9.4 WHAT IS EMBEDDED SYSTEM?

Embedded system (ES) is a system that has hardware circuit, with software embedded in it to get desired output. These systems contain microcontrollers or microprocessor with sensors, actuators as subsystems to make one complete system in itself designed for a specific dedicated application.

ESs are programmed to do a specific task but in general-purpose systems, any program can be run as per the choice of programmer. In ES, microcontroller or microprocessor act as heart of whole systems and extra circuitry is needed to achieve desired result as per application as presented by Fig. 9.2.

Software in ES is embedded in read only memory. It does not need alternative memory as in computer systems, as required by desired application.

The main components of ES are as follows and as shown in Figure 9.3:

(1) Hardware
(2) Software
(3) Real-time operating system (RTOS)

Hardware: Embedded hardware contains microcontroller, microprocessor, sensors, actuators, memories, input and output ports, timers and counters, etc.

Software: It is a set of rules used to perform single or multiple tasks. ES programming can be done via C, C++, assembly language, python, etc.

Real-time operating system: It is a system that processes real-time applications. It processes data without any delay and buffer. RTOS defines the way how the system works. During execution of instructions, it sets the rules to complete it successfully.

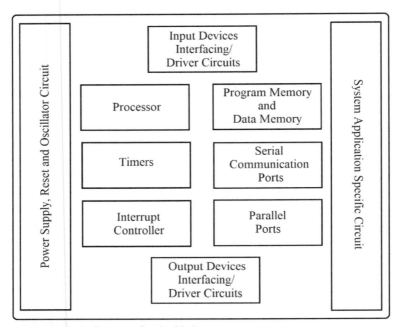

FIGURE 9.2 Block diagram of embedded system.

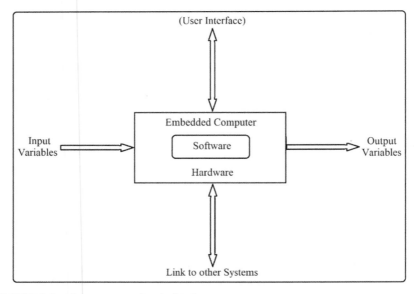

FIGURE 9.3 Components of embedded system.

9.4.1 TYPES OF EMBEDDED SYSTEMS

ESs are classified into following different types based on their performance, functional requirements, and performance of microcontrollers used as represented by Figure 9.4.

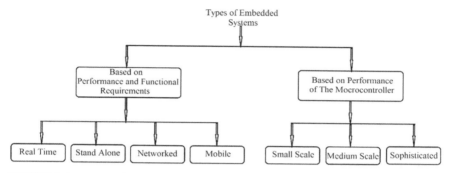

FIGURE 9.4 Types of embedded system.

ESs are categorized into four types based on functional needs:
- Stand-alone ESs
- Real-time ESs
- Networked ESs
- Mobile ESs

Stand-alone ESs: As its name defines, stand-alone ESs can work independently, they do not need any type of host system like computers. These types of systems receive input from I/O pins and can be analog or digital. Processing unit with the help of embedded programs processes the data and sends it to output devices to further display it on devices like LED and LCD or to control motors, etc. Such systems are simple in design and less complex. Examples are digital cameras, microwave ovens, and mp3 players.

Real-time ESs: Real-time ESs give output without any delay between specified times scheduled to complete one particular task. Time for accomplishing task can vary from application to application.

Real-time ESs are further classified into two systems: (1) soft real-time systems and (2) hard real-time systems.

Soft real-time ESs: In these systems, the quality of output is priority as compared to meet the deadline. So, completion time can vary but output produced by system after that interval is highly acceptable and useful; for instance, microwave oven—time taken by microwave to cook food can vary but it should be cooked well.

Hard real-time ESs: In such systems, time is the high priority than quality of output produced. Task must be accomplished within scheduled time period. So, output achieved after time has elapsed cannot be acceptable; for example, traffic light control system—in this system, it can be programmed to make light ON and OFF at defined time interval, otherwise, it can lead to casualties.

Network ESs: As it is clear from its name, such ESs needs some kind of network to get input from remote resources to produce desired output. Depending upon the distance of connection, network can be PAN (Personal Area Network), LAN (Local Area Network), MAN (Metropolitan Area Network), or WAN (Wide Area Network). Data sharing between devices like sensors on such networks can be done through wires or wireless media. In these networks, all devices can be connected to a server called Web server, data can be controlled and processed on this server using protocols by Web browser; for example, home automation system—where all sensors are connected to each other by TCP/IP protocol in a network. Another example is IoT device.

Mobile ESs: In every ES, now compactness is the main concern. These types of ESs are small in size and easy to integrate with mobile devices like phones, digital cameras, etc. In such systems, power and memory are the main constraints apart from size, cost, and wide variety.

Following are the three types of ESs based on microcontroller performance:

(1) Small-scale ESs
(2) Medium-scale ESs
(3) Large-scale ESs

Small-scale ESs: Small-scale ESs can be designed using 8- and 16-bit microcontrollers. So, system is simple in designing and less complex. These systems are used in simple applications like toys. Time requirements are not critical in such systems. These systems are often battery operated. They may or may not have operating systems. The coding language used is C and programming tools are assemblies, editor and cross assembler, and integrated development environment (IDE). Examples are fax machine, washing machine, and printer.

Medium-scale ESs: These systems are more complex in both hardware and software as compared to simple-scale ESs. They are generally built up using 16- and 32-bit microprocessor or microcontrollers. These types of processor used are DSP (Digital Signal Processing) with RISC (Reduced Instruction Set computer) and CISC (Complex instruction Set Computer) architecture. Programming tools used are C++, debugger, RTOS, simulator, and IDE. Medium-scale systems are fact as compared to small-scale systems, for example, automatic teller machine.

Large-scale ESs: These systems use 32/64-bit microprocessor or microcontrollers. These are more complex in terms of hardware and software as compared to small- and medium-scale systems. In such systems, speed is the major concern. Scalable or configurable processors and reprogrammable logic arrays are needed. Network routers, embedded Web server, and IP cameras are some examples of such systems.

9.4.2 CHARACTERISTICS OF AN EMBEDDED SYSTEM

Followings are some of the characteristics of ESs:

Application and domain specific: ESs should be domain specific or application specific. ESs designed for microwave control unit cannot be replaced with air conditioner (AC) control unit. Similarly, ESs designed for telecom cannot be replaced with consumer electronics systems.

Reactive and real time: Real-time ESs produce changes in output in response to changes in input signal is called reactive and if the system is responding to requests or tasks in known amount of time, it is called real-time ESs, for example, antilock systems and flight control systems.

Operates in harsh environment: The environment in which ESs are deployed can be dusty or one with high temperature zone or can be area that subject to more vibrations and shock. So, systems should withstand all these adverse effects.

Distributed: ESs can be the part of larger systems; for example, automatic vending machine consists of card reader or vending unit. Both are independent ESs but they work together to perform overall function to get output. Also, automatic teller machine consists of subunits like card reader unit that is responsible for reading and authenticating the cards, transaction unit, currency counting, printer unit for printing slips, etc.

Small size and weight: Compactness is a big concern now in every electronic circuit, so ESs should be light in weight and compact in size.

Power concerns: In electronic circuits, minimum heat dissipation is needed because high amount of heat dissipated requires more cooling.

Tightly constrained: All systems have some guidelines related to design, cost, power consumption, and size. So, ESs should be fast to process data without leading to higher latencies and can be fit easily on a single chip. These systems should be capable of processing real-time data with less power consumption.

Connected: ESs should have some peripherals to connect input and output devices.

9.4.3 ADVANTAGES OF EMBEDDED SYSTEMS

- Small in size
- Fast in operation
- Cheap in cost
- Require less power for operation
- More reliable systems
- Dedicated to one specific operation, so uses fewer resources

9.4.4 DISADVANTAGES OF EMBEDDED SYSTEMS

- Difficult in upgradation
- Limited in hardware
- Difficult to troubleshoot
- Difficult to integrate
- Hard to take backups

9.5 INTERNET OF THINGS (IOT)

The IoT concept came into existence in 1999 by RFID development community. With the growth of mobile devices, ESs, and cloud computing, it becomes point of attraction for the modern world. IoT consists of devices that have unique identities and are connected to Internet to sense and share information over IP networks. Data from all these connected devices are received on regular basis and used to initiate some actions for further decision-making. The Internet does not only mean network of computers, but it also includes devices like sensors, thermostats, irrigation pumps, Bluetooth-controlled devices, smartphones, vehicles, and cameras which are used to share information based on different protocols to achieve smart output like online monitoring, control, positioning, etc.

The embedded devices are the systems designed using microcontroller and microprocessors to build the autonomous computing system. These systems can share and process the data without using Internet but can be connected with it. However, these devices can connect through the Internet and able to communicate through it with other network devices. So, embedded IoT systems are microprocessor–microcontroller devices connected with Web server directly or indirectly. Three main components of IoT are as follows:

1. Embedded programming languages like C, C++, JAVA, Python, etc.
2. Network technology
3. Information technology

9.5.1 ROLE OF EMBEDDED SYSTEMS IN IOT

From last few years, tremendous growth has been observed in ES market with advent of IoT, to share data between connected devices on high speed. More and more devices are connected in real time for data sharing from remote locations due to IoT.

The IoT is basically defined as the network of devices consisting of processors, analog or digital sensors, actuators, and a set of instructions called programs used to design hardware circuits and develop printed circuit boards; further to connect to the Web, they need API (Application Programming Interface) and protocols to of embedded devices. This connected environment that allows user and technologies to communicate with each other either through wire or wireless media in PAN, LAN, or MAN networks has boosted up the interaction in the world digitally. These connected ESs are able to collect real-time data and process them by continuously changing behavior and the way they interact with our environment.

In our daily life, we came across various ESs such as AC, cars, vending machines, printers, microwave ovens, and washing machines, which can perform different and unique operations. For designing such systems to perform some specific operations, some tight constraints should be kept in mind like power consumption, size, cost, security, speed, reliability, etc., so designing IoT-based embedded hardware is not an easy task.

9.5.2 EMBEDDED IOT DESIGN ISSUES/CHALLENGES

While designing embedded IoT hardware, the following challenges are faced:

Lack of flexibility for running applications: To connect devices with each other, ESs should adopt to hardware architecture and new functions in real-time environment. So, system should adapt to new applications and technology, designers face the following problems:

1. Difficulty to adopt new environment
2. Problems in integration of new applications
3. Changes in software and hardware facilities
4. Low weight and less power consumption
5. Packaging and integration problems with small ICs

Security crisis in ES design: In ESs, security is the big concern of designers. In real-time environment, all devices should perform function securely if they are working in harsh conditions. To ensure security and robustness of such systems, designers use some cryptographic algorithms and security programs.

High power dissipation of ES: The main limitation of ESs is power dissipation. According to Moore's law, a number of transistors are increasing and size has to reduce, so it is the big challenge to integrate a high number of transistors on single chip with less power consumption. To reduce overall system power consumption, highly efficient architecture design is needed or system frequency can be increased.

Problems of testing an ES design: For efficient working of any system, it should be tested, verified, and validated properly that is another big concern. Hardware testing is done to check system performance and validate as per the requirement of system. Verification and validation are done to check whether product passes all quality standards.

Increased cost and time to market: Embedded hardware and software designers should bring the devices at appropriate time to the market with reduced cost.

9.5.3 APPLICATIONS OF IOT

- Healthcare
- Smart environment
- Transportation
- Home automation
- Smart surveillance
- Manufacturing applications

9.6 NEED OF ANTENNA IN EMBEDDED IOT

As we already know, IoT field brings multiple end devices that are connected with each other through radio signals like sensor, actuators, microcontrollers, processors, etc. so to transfer signals without using wires need antenna. The shape and choice of antenna used in circuit depends on application and frequency band used with other factors like transmission distance and strength of signal. As in 4G/5G communication networks that work on high-frequency bands with large bandwidths, IoT networks transmit and receive signals smartly in small chunks using different network topologies like star, mesh, bus, etc. Different types of antennas are used in IoT networks like wire antennas, patch, whip with omnidirectional radiation pattern. Various IoT development kits and boards like Arduino, Qualcomm use same type of antennas used in Bluetooth, Wi-Fi, and GPS communication applications. Antennas used in medical field for wearable applications are generally low-profile antennas. This concludes that the design of antennas varies from device to device to meet the requirements. Antennas in IoT applications can be classified as per frequency bands or on the basis of application. Table 9.1 shows some antennas used for IoT applications with frequency bands.

9.6.1 WHAT IS ANTENNA

Antenna is a passive device that consists of conductor material and is used to convert electrical energy into electromagnetic waves at transmitter side and from radio waves to currents signals on receiver side. When signals have to be transferred between two devices using wireless media, antenna is highly needed.

Microstrip antennas are gaining prominence in many applications in wireless communication devices due to numerous attractive features like small size, less weight, easy integration in circuits, low cost, and simple PCB fabrication. Instead of these advantages, microstrip patch antennas have various shortcomings in terms of gain and bandwidth.

TABLE 9.1 Wireless Standards with Frequency Bands.

IoT application	Wireless technology	Operational frequency
Home automation	Wi-Fi	2.4 GHz
	GPS	1575.42 MHz, 1227.6 MHz, 1176.45 MHz
Smart agriculture and cities	Zigbee	915 MHz, 2.4 GHz
	Bluetooth	2.4 GHz
	LoRa	433, 868, 916 MHz
	Sigfox	868, 902 MHz
Medical	MBAN/WBAN (IEEE 802.15.6)	400, 800, 900 MHz, 2.36 and 2.4 GHz
Avionics	WAIC	4.2 and 4.4 GHz

9.6.2 MICROSTRIP PATCH ANTENNA

Microstrip patch antenna consists of copper material on one side of dielectric substrate material and another side ground plan that also consists of conducting material. MPA (Microstrip Patch Antenna) patch can have different shapes like square, rectangular, triangular, circle, E shape, etc. Patch and ground are fetched using proper fabrication techniques on substrate as represented by Figure 9.5.

Role of Microstrip Patch Antenna for Embedded IoT Applications

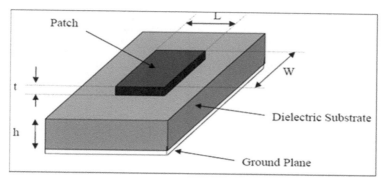

FIGURE 9.5 Microstrip patch antenna.

Dimensions of patch are considered with mathematical formulas. For rectangular shape, patch length L is taken between 0.33 times λ_0 and 0.5 λ_0, here λ_0 is the free-space wavelength. Patch thickness selected is very thin with thickness less than λ_0. Substrate height h is considered between 0.003 $\lambda_0 \leq h \leq 0.05 \lambda_0$. Dielectric constant of substrate used for antenna designing lies in range of 2.2–12, that is, $2.2 \leq \varepsilon_r \leq 12$.

For antenna designing, to find its dimensions, transmission line theory can be considered. To design rectangular microstrip patch antenna resonating frequency F_r, substrate dielectric constant ε_r, and height of substrate h should be known. So, all antenna parameters can be calculated using the following design expressions:

$$\lambda = \frac{C}{F_r} \tag{9.1}$$

where λ is the wavelength and C is the velocity of light and F_r is the resonating frequency.

Microstrip antenna width (W) can be calculated using the following expressions:

$$W = \frac{C}{2Fr}\sqrt{\frac{2}{\varepsilon r + 1}} \tag{9.2}$$

where C is the speed of light (3×10^8 m/s)

Eligible effective dielectric constant of the substrate is calculated using the following expressions:

$$\varepsilon_{reff} = \frac{\varepsilon r+1}{2} \frac{\varepsilon r-1}{2}\left(1+12\frac{h}{w}\right)^{-1/2} \quad (9.3)$$

where h is the thickness or height of the substrate and w is the width of the antenna. Effective length of the antenna at resonant frequency is calculated as:

$$L_{eff} = \frac{C}{2Fr\sqrt{\varepsilon reff}} \quad (9.4)$$

After considering fringing effect, antenna length extension is calculated as:

$$\Delta L = 0.412h\left[\frac{\varepsilon reff+0.3}{\varepsilon reff-0.258} \frac{\frac{W}{h}+0.264}{\frac{W}{h}+0.8}\right] \quad (9.5)$$

The total length of antenna is calculated as:

$$L = L_{eff} - 2\Delta L \quad (9.6)$$

Substrate width and length is found as:

$$L_g = L + 6h \text{ and } W_g = W + 6h \quad (9.7)$$

Height of substrate:

$$h = \frac{0.0606\lambda}{\sqrt{\varepsilon r}} \quad (9.8)$$

Feed line is given as:

$$L_f = \lambda_g/4$$

$$\lambda g = \frac{\lambda}{\sqrt{\varepsilon eff}} \quad (9.9)$$

Antenna efficiency is calculated as:

$$\eta = (gain/directivity) \times 100 \quad (9.10)$$

Radiation box calculation is done as follows:

Axis position = $(-\lambda_g/6) + (-\lambda_g/6) + (-\lambda_g/6)$

$$\text{Length} = (-\lambda_g/6) + (-\lambda_g/6) + L_g$$
$$\text{Width} = (-\lambda_g/6) + (-\lambda_g/6) + W_g$$
$$\text{Height} = (-\lambda_g)/6 + (-\lambda_g)/6 + h_g \tag{9.11}$$

9.6.3 MICROSTRIP ANTENNA FEED TECHNIQUES

Microstrip patch antenna is fed using different feeding methods like microstrip feed line, coaxial probe, aperture coupling, and proximity coupling feed methods. Each technique has its own advantages and disadvantages as discussed next.

(1) Microstrip feed line: The microstrip feed line is used mostly in MPA antenna fabrication due to simple in design and easy in construction. Antenna patch impedance matching is done with feed line impedance with proper calculations and can be varied by shifting the feed position. Microstrip antennas using this feed method generally provides less bandwidth as given in Figure 9.6.

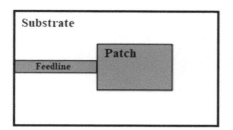 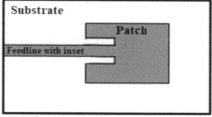

FIGURE 9.6 Geometry of microstrip feed line: (a) directly feed (b) inset feed.

(2) Coaxial probe: Coaxial feed or probe feed technique is used to feed antenna patch from ground plan through substrate. Coaxial feed line consists of two conductors: inner and outer. Inner conductor is attached to antenna patch and outer is connected to ground plane. Impedance matching is good and easy to fabricate too. It also has low bandwidth and not used if substrate thickness is more as illustrated by Figure 9.7.

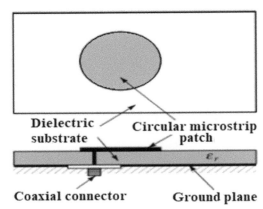

FIGURE 9.7 Geometry of coaxial probe feed patch antenna.

(3) Aperture coupling technique: This feeding technique uses two substrates and ground plan is sandwich between substrates. Microstrip feed line is on bottom substrate and is coupled with patch on top substrate through a small aperture in the ground plan. This method is mainly used in antenna arrays but it is difficult to fabricate due to two substrate layers that also increase antenna thickness as given in Figure 9.8.

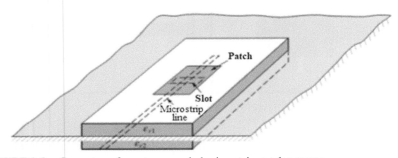

FIGURE 9.8 Geometry of aperture coupled microstrip patch antenna.

(4) Proximity coupling feed antenna: This method is used with thick substrates. It also uses two substrates with microstrip feed line in between. Its another name is electromagnetic coupling. Radiating patch is on the top of upper substrate. This feeding method provides high bandwidth achievement compared to other techniques but it is difficult to fabricate because both substrate materials need perfect alignment as given in Figure 9.9.

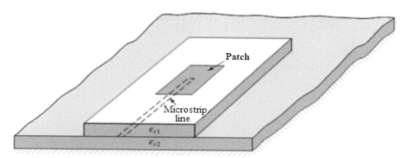

FIGURE 9.9 Geometry of proximity coupled microstrip patch antenna.

9.6.4 ADVANTAGES AND DISADVANTAGES OF PATCH ANTENNAS

Advantages of microstrip patch antennas are given as follows:

- Light in weight
- Low profile configuration
- Fabrication cost is less, so fabricated in bulk
- Can be linearly or circularly polarized
- Easy integration with microwave circuits
- Able to generate multiple resonating bands

Instead of having these advantages, microstrip patch antennas have several disadvantages as listed here:

- Narrow impedance bandwidth
- Low working efficiency
- Less gain
- Extra radiation generated from feeds and junctions
- Less power handling capacity

9.6.5 APPLICATIONS OF MICROSTRIP ANTENNAS

- Aerospace vehicles, including high-performance aircraft, spacecraft, satellites, and missiles
- Mobile radios, phones, and pagers
- Wireless applications like RFID, WiMAX, and Wi-Fi
- IoT applications

9.6.6 ANTENNA SIMULATION SOFTWARE USED

- HFSS
- CST microwave studio
- IE3D

9.7 CONCLUSION

IoT is the fastest-growing field in Industrial Revolution to handle the SATA (Serial Advanced technology Attachment) in more smartly. To build smart systems, to share data on remote locations, and to connect huge devices with each other, the concept of IoT is used with integration to ESs. In agriculture, health, transportation, and communication applications, IoT network can be built through different devices like sensors, switches, processing devices like microprocessors and microcontrollers, and some programming languages. With this tremendous growth in technology in all the fields, IoT boosts up the need for wireless technologies. From small sensors to Internet gateways, information exchange took place through radio-frequency signals. To share data through wireless medium, efficient microstrip patch antennas are highly regarded at transmitter and receiver sides to make wireless communication between devices to operate them on dedicated frequency bands without interference, with less power consumption and to transfer data at high data rate over long distances.

KEYWORDS

- embedded
- IoT
- microstrip patch antenna
- microcontroller
- microprocessor

REFERENCES

1. Katoch, S.; Jotwani, H.; Pani, S.; Rajawat, A. In *A Compact Dual Band Antenna for IOT Applications*, IEEE International Conference on Green Computing and Internet of Things (ICGCIoT), 2015; pp 1594–1597.

2. Vikram, N. In *Design of ISM Band RFID Reader Antenna for IoT Applications*, International Conference on Wireless Communications, Signal Processing and Networking (WiSPNET), 2016; pp 1818–1821.
3. Asadallah, F. A.; Costantine, J.; Tawk, Y.; Lizzi, L.; Ferrero, F.; Christodoulou, C. G. In *A Digitally Tuned Reconfigurable Patch Antenna for IoT Devices*, IEEE International Symposium on Antenna and Propagation & USNC/URSI National Radio Science Meeting, 2017; pp 917–918.
4. Vyshnavi Das S. K. In *Design of Triple Starfish Shaped Microstrip Patch Antenna for IoT Applications*, IEEE International Conference on Circuits and Systems (ICCS), 2017; pp 182–185.
5. Malek, N. A. In *Beamforming Capability of a Cross-Slotted Ring Patch Antenna Array at 2.4 GHz Wireless Applications*, IEEE 7th International Conference on Computer and Communication Engineering (ICCCE), 2018; pp 119–123.
6. Kumar, P. In *Design and Analysis of Multiple Bands Spider Web Shaped Circular Patch Antenna for IoT Application*, IEEE International Conference on Power Electronics (IICPE), 2018.
7. Sabban, A. In *Small New Wearable Antennas for IOT, Medical and Sport Applications*, IEEE 13th European Conference on Antennas and Propagation (EuCAP 2019), 2019.

CHAPTER 10

Sensible Vision Using Image Processing

KANWALJEET SINGH[1*], SHIVANG TYAGI[1], and BALJEET KAUR[2]

[1]*School of Electronics and Electrical Engineering, Lovely Professional University, Phagwara, India*

[2]*GNE College of Engineering, Ludhiana, India*

*Corresponding author. E-mail: kamal1997@gmail.com

ABSTRACT

Millions of humans stay in this world unable to be employed because of impairment in vision. Still they can generate alternative methods to deal with day-to-day routines and additionally suffer from navigation difficulties as well as social awkwardness. This advance research is on obstacle detection and recognition. The existing system is unable to multitask, but an advance research system will be able to multitask like object detection and object recognition, navigation, audible surrounding voice, calculation of object distance, and many more. With sensible vision, a blind person is able to identify how far an object is and how to reach it. Sensible vision can work as a timer or alarm system. It helps the blind person to know the face expression of the person with whom they are talking. It is more stable than before. Sensible vision can give fast response with the help of some algorithm. Sensible vision has Bluetooth system by which we can make connection between system and any kind of sound system. This chapter is going to explore a real-time object detection for the purpose of alerting the user about surrounding objects and their positions by using sound.

10.1 INTRODUCTION

A lot of humans stay in this world unable to be employed because of impairment in vision. They are generating alternative methods to come out with the optimal solution for their daily routines and facing obstacles to come up. This advanced research is on obstacle detection and recognition. The existing system is unable to multitask, but an advance research system will be able to multitask like object detection and object recognition, navigation, audible surrounding voice, and calculation of object distance.[1] Sensible vision can give fast response with the help of a specific set of algorithms.[2] Sensible vision has Bluetooth feature by which we can make connection between the system and any kind of sound system. On the contrary, in case of unfamiliar environment, it is difficult for them to identify a place. In addition, they cannot analyze that whether the person with whom they are talking to is interested to talk to them or not.[3] Deep learning can process or observe the things similar to a human brain.[4] The structure following deep leaning is called an artificial neural network. In machine learning, we have to fill the features manually, and the feature can be the size or shape of an object. In deep learning, the features are picked out by the neural network without human intervention.[5] Therefore, it can predict that whether an email spam is a good email that needs to be delivered in inbox.[6] Moreover, blind people cannot find out whether a person is talking to them or someone else who is standing near them during the conversation.[7] Therefore, according to previous research, this system is based on the use of new applied sciences to enhance the mobility of visually impaired humans. With sensible vision, a blind person is able to know that how far an object is and how to reach it. Sensible vision can also work as a timer or alarm system. It helps the blind person to know the facial expression of the person with whom they talk. It is more stable than before. Camera is used to detect the object and recognize it.[8]

This chapter is explores a real-time object detection system for the purpose of alerting the user about surrounding objects and sounds by using audio signals. Girshick and Ross gave their words related works on sensory substitution and assisted the development of products using computer vision for blind people and the exploration of 3D sound.[9] This introduces distinct aspects of our prototype.

In this chapter, we want to know the behavior of a person, if the person wants to talk or not.

10.2 SYSTEM DESCRIPTION

The core of the system is a portable module where there is a microcontroller that handles communication with the camera and sends the useful information to the user via Bluetooth. It also takes surrounding audio information through a microphone and gives the information to Bluetooth. The module has inbuilt 3.5-mm stereo-type audio jack (TRS) to give input to the headphone.

Microcontroller: The device has installed one microcontroller that gets the information with the camera to detect the object by processing an algorithm. Object detection is the constituent of object recognition. It finds the instance object from the real-time video processing. Else it will take the information by face recognition. This device has an installed microphone that gives the output as audio information.

First, the objects are recognized by the deep learning and machine learning algorithms. Then the algorithm converts the information into audio. After modification, information is sent to the user who takes the action.

Deep learning: Have you ever wondered how Google translates an entire webpage in a different language in a matter of seconds or our phone galleries group images based on their locations? All of these are an outcome of deep learning. But what is the meaning of deep learning for object detection? Deep learning is a subdivision of machine learning that in turn is a subdivision of artificial intelligence (AI). AI is a machine that mimics human behavior. Machine learning is the way to train AI through applying algorithms on data. Deep learning can process or observe the things as a human brain process.[4] The form that is following deep leaning is called an artificial neural network. In machine learning, we have to fill the features manually, and the feature can be the size or shape of an object. In deep learning, the attribute is distinguished by the neural network without human intervention.[5] That kind of freedom arises at the cost of having a much big volume of data to train our machine.

Neural networks: Neural network is the network of neurons. Any data has three different styles of word written like letter "A." If "A" is not written identically, the human brain can easily recognize the letter. When it comes to computers, they have to recognize them with the help of deep learning. The neural network trained to identify handwritten digits, and each number is presented as an image of X times X pixels. Then this amounts to a total of Y pixels. Neurons, the core entity of neural network, are where the data processing takes place. Each of the Y pixels is sustained to a neuron in the initial layer of our neural network. This forms the input layer. The output layer with each neuron represents a digit with the hidden layer existing between them, and the information is transformed from one layer to another over connecting channels. Each of these has a value attached to it and hence is called a channel having some weight. Each and every neuron has a distinctive number related to it, which is known as bias. It is applied to a function known as the activation function. If the neurons get activated, the result of the activation function governs. By the following layers, every neuron information passes. This continues till the second last layer. The one neuron actuated in the output layer relates to the input digit. The weights and bias are continuously adjusted to produce as well-trained networks. Therefore, in deep learning, neural network requires a massive volume of data to train.

$$X \times X = Y$$

Machine learning: Machine learning is a subset of AI. Machine learning and AI are the dissections of computer science. Machine learning is the narrowly related to data mining. Data mining is the evolution of data science that has been present quite for some time, for example, spam emails. Among the emails, some of the emails are sent into the inbox and some are into the spam box. It happens because of machine learning. There is huge chunk of data, and the program and algorithm are planned like that. Therefore, it can predict that whether an email spam is a good enough to be delivered to the inbox.[6] For object detection, we need to take some bunch of data and store it. Then with that the device will be trained.

Display: The device consists of a display to show the input that has been taken and also the output that is required. It recognizes the available neighboring Bluetooth device to connect the two devices. Also, it shows the distance between the objects and device.

Note: Display is required for mainly two reasons:

1. for repairing purpose,
2. if the user needs any kind of help from the other person to tell them the processing and output showing in the display.

Camera: The live streaming can be used to recognize object in the actual world. Camera broadcast can detect object and use them as input to perform an observable search. The device has inbuilt camera module that is the most important part to perform task. Camera is used to detect the object and recognize it.[8] Camera gives the information to microcontroller, and then by using deep learning and machine learning, it completes the task, for example, distance calculation, object detection, object recognition.

Microphone: The device has installed one microphone to provide the surrounding audio data to the user. Microphone is needed if the user is using earphones, headphone, or any kind of wearable device to listen the output. Therefore, the microphone provides the surrounding audio at the knee-high volume. Nowadays, we hear a lot of news of accidents as people use a lot of devices like earphones and headphones, because of which they are unable to catch up with the surrounding sound. So, for that kind of problem, the device has a microphone inside it to present the sound of the surrounding environment at the low volume (audio). Microphone is used for the protective purpose only.

Bluetooth: The device has an installed Bluetooth module to make communication between the Bluetooth device and user device. Bluetooth is a wireless technology that enables you to create a wireless connection between the device and the module. In this case, module uses Bluetooth for wireless audio transmission. The motive behind to the use Bluetooth module is to enable the user to connect the device to their home loudspeakers and ear pots. One of the most widely recognized uses for Bluetooth is to associate your cell phone to remote speakers or earphones. The upside of this kind of earphones is that you do not have to stress over links or wires getting tangled or pulled, one reason why Bluetooth is especially valuable for sports earphones. You can likewise find a huge number of little and amazing Bluetooth speakers to fit all your needs.

Audio jack: The module has installed audio jack to make communication between module and user. Module has 3.5-mm stereo-type audio jack to make wired connection between user and module. The motive of providing audio jack is if the user is moving outdoor then the user can also use wired earphones. The 3.5-mm stereo jack is used because numerous people use this type of earphones.

3D sound device: Whenever we use the sound device and we listen to the left or the right sound configuration, the audio from the left speaker arrives at both our ears and is summed with the input from the right speaker. But as we listen to the same sound using earphones, then the left ear only uses the left channel and right ear uses the right channel.

Raspberry Pi: The Raspberry Pi was constructed in light of the objective of training. This ultralittle PC was intended to be little and modest with the goal that schools could without much of a stretch manage the cost of them so as to train students on computers. This is extraordinary for two reasons: the first is that it gives amazingly modest access to a PC, and second, it is an incredible device for getting familiar with PCs (understudy or not)! Raspberry Pi is a great system. This is likely the best component of the Raspberry Pi, yet in all likelihood is one of the least examined. The social order that uses the Raspberry Pi is the most pleasing one that I have seen over the tech extent. One of the benefits of the informational piece of this contraption is that each undertaking someone does with it is incredibly all around documented with a tiny bit at a time heading and normally consolidates pictures. The conversations for the system run various zones and every level of ability.

10.2.1 HOW DOES THE RASPBERRY PI WORK?

Here is the method by which it works: an SD card installed into the opening on the board goes about as the hard drive for the Raspberry Pi. It is powered by USB, and the video yield can be seen on a standard RCA TV set, a reliable present-day screen, or even a TV using the HDMI (High-Definition Multimedia Interface) port. This gives everyone the fundamental limits of an ordinary PC. It uses low power of around 3 W.

10.2.2 BRIEF DESCRIPTION OF RASPBERRY PI

Processor: BCM2835 SoC module is a Broadcom processor used in Raspberry Pi. It has an ARM processor that is an advanced RISC (Reduced Instruction Set Computer) machine.

The Broadcom SoC is just like a chip that is used in the Raspberry Pi. While working at a clock speed of 700 MHz by default, the Raspberry Pi gives an authentic execution by and large similar to the 0.041 GFLOPS (billions of Floating-Point Operations per Second). On the CPU level, it resembles a clock speed of 300 MHz Pentium II (1997–1999), yet the GPU, regardless, gives 1 and 1.5 Gpixel/s or 24 GFLOPS of all-around valuable figure and the representations limits of the Raspberry Pi are commonly proportionate to the level of execution of the Xbox of 2001. The Raspberry Pi chip works at 700 MHz normally, and so that it does not get too much hot, it requires a heat sink or remarkable cooling.

Power source: The Raspberry Pi is a gadget that runs on 700 mA or 3 W. The power supply is through a micro-USB charger or the GPIO (General-Purpose Input/Output) header. To drive the Pi, any considerable cell phone charger can be used.

SD card: In Pi, no onboard storage is available. The working structure is stacked on a SD card that is implanted in the SD card slot on the Raspberry Pi. Operating system for the Raspberry Pi is installed in the SD card only with the help of a card reader.

GPIO: It is a general pin which is on an integrated circuit and can be used as input or output pin as per the requirement by the user.

Figure 10.1 describes that the GPIO pins have no specific reason characterized and go unused as a matter of course. The thought is that occasionally the framework planner fabricating a full framework that uses the chip may think that it is helpful to have a bunch of extra computerized control lines, and having these accessible from the chip can spare the problem of organizing extra hardware to give them.

GPIO ability may include the following:

- general-purpose input/output pins, as name suggests, it can be used as input or output.
- general-purpose input/output pins can be activated/deactivated,
- input data are understandable (typically high/True = 1, low/False = 0),
- output data can be writable as well as readable, and
- IRQs (Indian Register Quality Systems) sometimes can be used as input data.

The present-generation Raspberry Pi development board has a 26-pin 2.54-mm (100 mil) development header, set apart as P1, masterminded in a 2 × 13 strip. They give eight GPIO pins in addition to access to inter integrated circuit, serial peripheral interface, UART), just as +3 V, +5 V, and GND (Ground (0 V)) supply lines. The primary segment pin is pin 1 and on the base line. The detailed description follows Figure 10.1.

	Pin No.		
3.3V	1	2	5V
GPIO2	3	4	5V
GPIO3	5	6	GND
GPIO4	7	8	GPIO14
GND	9	10	GPIO15
GPIO17	11	12	GPIO18
GPIO27	13	14	GND
GPIO22	15	16	GPIO23
3.3V	17	18	GPIO24
GPIO10	19	20	GND
GPIO9	21	22	GPIO25
GPIO11	23	24	GPIO8
GND	25	26	GPIO7
DNC	27	28	DNC
GPIO5	29	30	GND
GPIO6	31	32	GPIO12
GPIO13	33	34	GND
GPIO19	35	36	GPIO16
GPIO26	37	38	GPIO20
GND	39	40	GPIO21

FIGURE 10.1 GPIO pins of Raspberry Pi.

DSI connector: The connector uses the mobile industry processor interface. Alliance is planned for diminishing the expense of show controllers in a cell phone. LCD and comparative presentation advancements are focused by DSI. It characterizes a sequential transport and a correspondence convention between the host (wellspring of the picture information) and the gadget (goal of the picture information).

DSI connector is used for good LCD screen, despite the fact that it might require extra drivers to drive the presentation.

Audio jack: A good-quality 3.5-mm connector named TRS is installed on the Raspberry Pi for stereo sound yield. Any link of 3.5-mm sound can be associated straightforwardly.

In spite of the fact that this jack cannot be utilized for taking sound information, USB sound card or USB microphones can be utilized.

10.3 METHODS

It is a real-time working project. Therefore, it has three modes to operate according to the situation.

Sensible Vision Using Image Processing

Home: This mode inspired by day-by-day routine at home for a blind person. First, we did a survey of regular problems that a blind person faces in the house. Then we get to know that they hardly identify the time in the morning, and they are unable to get where the object is placed, which they needed. To make them comfortable, we made a system that works as an alarm, identify the object for them and give them instruction to reach there.

The system gives the surrounding information and reminds them at which time what work they should do. Proper explanation of the project is given in Figure 10.2 in flowchart.

Flowchart

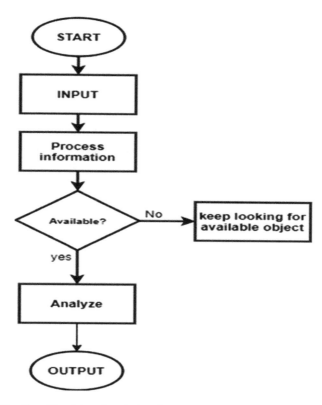

FIGURE 10.2 Sensible vision description through flowchart.

START: The device will start by pressing a push button. As it has three mode, it has three push buttons to switch the mode.

INPUT: There are two types of INPUTs that are taken by the device. The device has a camera that takes real-time video as INPUT. Object detection and recognition will use the real-time video streaming. The device has a microphone that takes the surrounding audio as INPUT. The surrounding audio will be used for protective purpose.

Process information: The device has a home mode to detect the surrounding environment of home. Therefore, while processing information, the device will search for the object and recognize the surrounding sounds.

Available: By following the flowchart, if any household appliance and surrounding audio is detected by the "process information," then it will go for further processes. If the device does not detect any household appliances, it will keep looking for available object as well as available surrounding audio.

Analyze: If the device finds out any object then by applying algorithms, the device will recognize the object then give the OUTPUT. In this mode, the surrounding environment will be household appliances. The object detection and object recognition algorithm used reads the INPUT taken by camera.

OUTPUT: The device will present the recognized object and surrounding audio as OUTPUT. The OUTPUT can be follows:

- sugar at 3 m of distance on the 90° of device,
- milk at 2 m of distance on the 45° of device,
- black mug is full of milk,
- refrigerator is at the 2 m of distance in 90° of the device, and
- surrounding audio.

Walking: The reason to make this mode is that nowadays we hear a lot of accident news on roads. Therefore, we find that what problem a blind person faces on road. Then we get to know that they do not know what is going on in the way they are walking. Therefore, for this kind of problem, we made this mode that enables the user to give them a signal in the form of sound. By using live streaming, the module is able to give them a clear way to walk and give them instruction by object detection and recognition. With the help of a map, the system is able to identify the way where the person wants to go.

Talking: The reason to make this mode is to know the face expression of whom the blind person is talking to. While talking to someone, a blind person wants to know whether the person to whom they are talking likes the conversation or not. Therefore, by analyzing that problem, we made this mode that is able to identify the person's face expression to whom a blind person is talking.

For example, neutral, happy, sad, fear, contempt, surprise, anger, and disgust.

For this purpose, you must know the basic python programming. Therefore, this mode is to be made up with the help of Python, OpenCV, and a Face Database.

Python: Python is a general-purpose language that is open source. It is procedural, object oriented, and functional language. It is easy to interface with C, C++, Objective-C, Java, and Fortran. It has incredible relative environment.

OpenCV: The aim of OpenCV is to do real-time computer vision. It is an open-source package/library. The library is a platform—it can support Python, Java, C++, etc. It was mainly introduced by Intel. It is free for use and is open-source BSD license. The OpenCV library is the most widely used library for video detection, motion detection, video recognition, image recognition, and even deep learning facial recognition.

Database: To organize the database in the directory, create two folders known as "source emotion" and "source images" and extract the content in the .txt files (S005, S010, etc.) in a folder known as "source emotion." Put the folder comprising the image in a folder known as "source image." One more folder is needed to be created as "source set" to house our sorted emotion images. Under this folder, we need to create folders for the different emotion labels ("sad," "happy," "angry," and so on).

How does it work?

This mode is impaired by GNU 3.0 open-source license. And it is free to change and reallocate the code.

10.4 RESULT

As we have three modes to operate the system, the system has three results.

Home: In the home mode, blind persons are looking for time and object detection and recognition. We have several things in our houses to recognize like water bottle, table, person, books, etc. Figures 10.3 and 10.4 describe the testing of object detection algorithm.

FIGURE 10.3 Testing of the program using cell phone detection.

FIGURE 10.4 Testing of the program using bottle detection.

Sensible Vision Using Image Processing

Walk: In the walk mode, blind persons are looking for several objects like car, traffic light, person, obstacles, etc. Figure 10.5 shows that different objects have been detected at the same time.

FIGURE 10.5 Detection of different objects in the image.

Talking: In the talking mode, the blind person wants to know while talking what is the expression of other the person, which is shown in Figures 10.6 and 10.7.

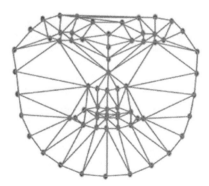

FIGURE 10.6 Human face expression.

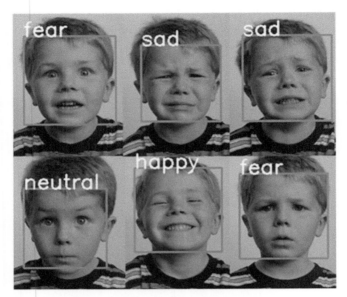

FIGURE 10.7 Different patterns of human face.

10.5 CONCLUSION

The research is about to do multitask to give a helping hand to blind persons. It has ability to sense the surrounding environment, make the data, and give that data to the user in the audio structure. It works based on machine and deep learning algorithms. It eases the way of living in this world for a blind person. As the name, sensible vision suggests the working of the project itself. It detects and recognizes the obstacles as well as face recognition in real time.

KEYWORDS

- **sensible vision**
- **recognition**
- **real-time object detection**
- **face expression**

REFERENCES

1. Wu, Y.; Yang, M.-H.; Lim, J.; Wu, Y. In *Online Object Tracking: A Benchmark*, Proceedings of the IEEE Conference on Computer Vision and Pattern Recognition, 2013; pp 2411–2418.
2. Canny, J. A Computational Approach to Edge Detection. *IEEE Trans. Mach. Intell. Pattern Anal.* **1986**, *6*, 679–698.
3. Schlecht, J.; Björn, O. In *Contour-Based Object Detection*, BMVC, 2011; pp 1–9.
4. Otsu, N. A Threshold Selection Method from Gray-Level Histograms. *IEEE Trans. Cybern. Syst. Man* **1979**, *9* (1), 62–66.
5. Torralba, A.; Quattoni, A. In *Recognizing Indoor Scenes*, 2009 IEEE Conference on Pattern Recognition and Computer Vision; IEEE, 2009; pp 413–420.
6. Betsworth, L.; Jones, M.; Srivastava, S.; Rajput, N. In *Audvert: Using Spatial Audio to Gain a Sense of Place*, Human-Computer Interaction in IFIP Conference, Berlin; Springer, Heidelberg, 2013; pp 455–462.
7. Xiao, J.; Ramdath, K.; Losilevish, M.; Sigh, S.; Tsakas, A. In *A Low Cost Outdoor Assistive Navigation System for Blind People*, Industrial Electronics and Applications in 2013 IEEE 8th Conference (ICIEA); IEEE, 2013; pp 828–833.
8. Ren, S.; He, K.; Girshick, R.; Sun, J. In *Faster R-CNN: Towards Real-Time Object Detection with Region Proposal Networks*, Advances the Neural Information Processing Systems, 2015; pp 91–99.
9. Girshick, R. In *Fast R-CNN*, Proceedings of the IEEE International Conference on Computer Vision, 2015; pp 1440–1448.

CHAPTER 11

Detection of Blocking Artifacts in JPEG Compressed Images at Low Bit Rate for IoT Applications

ANUDEEP GANDAM[1*], JAGROOP SINGH SIDHU[2], and MANWINDER SINGH[3]

[1]*Department of Electronic and Communication Engineering, IKG-Punjab Technical University, Jalandhar, Punjab, India*

[2]*Department of Electronic and Communication Engineering, DAVIET, Jalandhar, Punjab, India*

[3]*Department of Electronic and Communication Engineering, Lovely Professional University, Phagwara, Jalandhar, Punjab, India*

*Corresponding author. E-mail: gandam.anu@gmail.com

ABSTRACT

In the era of multimedia applications and the Internet of things, large bandwidth (a limited resource) is required for exchange of data to control various devices over the Internet and hence data compression is highly desirable. During compression, various artifacts may occur. This chapter focuses on an efficient technique of artifact detection. Blocking artifacts are predominantly visible during the process of block discrete cosine transform–based compression and this is the main cause of image degradation. A method to detect these unwanted effects without the removal of intuitive features of an image has been presented in this chapter. In the proposed work, a modified model based on the assumption of the large variation of pixel values of adjacent pixels across the block boundary in comparison to pixels away from the boundary is designed. The experimental results show that the proposed method outperforms while detecting blocking artifacts as compared to other postprocessing methods.

11.1 INTRODUCTION

In recent scenario, high data transmission is the need of hour, specifically in multimedia and the Internet of things (IoT) applications, which requires large bandwidth. The main constraint in transmission of data is basically nonavailability of bandwidth. In image and video data transmission, compression of data is required to mitigate the problem of large bandwidth. Earlier compression was done by just implementing JPEG, JPEG-2000 for images, and MPEG for video, which itself produced lots of artifacts. Nowadays, a lot of development has been done in this area. Image and video compressions have gained astounding importance in the modern era of communication. The JPEG image coding standard and standards of video compression have relied on block discrete cosine transform (DCT) (BDCT) as the first and foremost step. However, the major drawback is that it introduces artifacts in the compressed images that are clearly visible at low bit rates.[1–5] Detection and removal of these artifacts is the need of the hour. Figure 11.1 shows four JPEG coded images at different bits per pixel (BPP). It can be clearly seen that the blocking artifacts become more visible at higher compression ratio.[6–7]

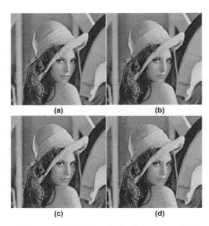

FIGURE 11.1 Lena at different BPP: (a) original Lena at BPP = 0.999, (b) Lena at BPP = 0.86, (c) Lena at BPP = 0.45, and (d) Lena at BPP = 0.37.

11.2 RELATED WORK

Wang et al. proposed a blind measurement method in which the intensity of blocking artifacts can be measured in the blocky images without using any data from the original image. Blocking artifacts' estimation is proficient in assessing the energy of blocky signal.[7] Liu et al. proposed a direct capacity to supplant the progression function.[8] This thought is straightforward but it has complex calculation process for recognizing blocking artifacts. Additionally, it cannot give great results when they ascertained visual estimation. Luo et al. joined edge detection technique with blocking artifacts. It replaced the calculated values after averaging with the calculated transform coefficients. This method had fewer calculation complexities and gave better results. However, it had a major drawback that the method was unable to justify the reason for the adopted technique.[9] The difficulty of Wang et al. is to expect that the distinction of the pixel esteem in piece limit is caused just by blocking artifacts. This suspicion diminishes calculation of many-sided quality yet the deliberate esteem does not affirm the truth for the two adjoining hinders with a slow change in pixel value.[10] In the technique proposed by Park et al.[11] and Singh et al.,[12–14] the variety of pixel esteem crosswise over piece limit was demonstrated as a direct capacity and exponential capacity separately. This technique is not exact for the nearby squares when a minor adjustment of qualities happens in pixels over the piece limit.

11.3 PROPOSED WORK

An attempt has been made to detect the blocking artifacts created due to image compression and additionally enhance the approach exhibited in Ref. [14]. For the better elaboration of block-based transform coding, an image is made up of nonoverlapping $N \times N$ blocks. Let "x" and "y" be the neighboring blocks of the image, which are horizontally adjacent to each other and have the block size of 8×8.

Vertically adjacent blocks can also be modeled in the same way. In the proposed method as shown in Figure 11.2, right sides of 8×8 block "x" and left sides of 8×8 block "y" seem to form a new block of 8×8 named "z." If any sort of artifacts are visible in the marked areas, the pixel values of the newly formed block "z" will change in an abrupt manner.

An extensive work has been done to reduce the so formed artifacts by operating on the block "z."

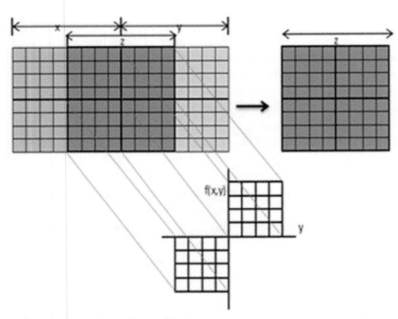

FIGURE 11.2 Formation of a new block.

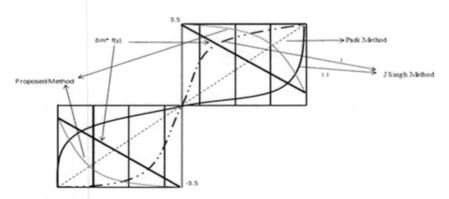

FIGURE 11.3 Measurement of blocking artifacts.

Detection of Blocking Artifacts in JPEG Compressed Images

An efficient postprocessing method is required to eliminate the artifacts along with maintaining the features of the image. The artifacts are visible across the boundaries both in horizontal and vertical direction (Fig. 11.3).

11.3.1 EXISTING METHOD

The method demonstrated by Park et al.[11] is based on the assumption that the pixel values do not change abruptly across the block boundaries of the processed image. This gradual change can be modeled as a two-dimensional (2D) linear function $m(y, x)$ given by

$$m(y, x) = m(x) = x - (M-1)/2 \qquad (11.1)$$

where (y, x) are the values of pixels, M is the size of the block, and proposed work is worked on for the size of 8. The values of $m(y, x)$ are antisymmetric in a horizontal direction but constant in vertical direction.

The method demonstrated by Singh et al.[14] gave a new function to process the image as shown in Figure 11.3, given by

$$f(x, y) = f(y) = \frac{y - [(M-1)/2 \pm 3.5]}{2} \qquad (11.2)$$

where $f(x, y)$ is the pixel values and M ($M = 8$) is no. of blocks of an image.

11.3.2 PROPOSED DETECTION METHOD

The difference between the values of the pixel along the block boundaries can be considered as the whole summation of blocking artifacts. Consider that the change in pixel values by a large magnitude instead of varying gradually as shown in Figure 11.3.

Then, the change in the value of "z" changes w.r.t. $f(x, y)$, the values of which are deceitful between the linear-valued function and a step function.

$$f(x, y) = f(y) = \frac{M-24}{2} - \frac{y}{2} \quad \text{for } y = 0, 1, 2, 3 \qquad (11.3)$$

$$f(x, y) = f(y) = \frac{M-1}{2} - \frac{y-4}{2} \quad \text{for } y = 4, 5, 6, 7 \qquad (11.4)$$

Equations 11.3 and 11.4 form a 2D block "b" that is antisymmetrical in the horizontal direction but constant in a vertical direction.[11] This block "b" is simply obtained by designing the vector "l" such that "b" equals to

$$b = \begin{vmatrix} l_0 & l_1 & l_2 & l_3 & l_4 & l_5 & l_6 & l_7 \\ l_0 & l_1 & l_2 & l_3 & l_4 & l_5 & l_6 & l_7 \\ l_0 & l_1 & l_2 & l_3 & l_4 & l_5 & l_6 & l_7 \\ l_0 & l_1 & l_2 & l_3 & l_4 & l_5 & l_6 & l_7 \\ l_0 & l_1 & l_2 & l_3 & l_4 & l_5 & l_6 & l_7 \\ l_0 & l_1 & l_2 & l_3 & l_4 & l_5 & l_6 & l_7 \\ l_0 & l_1 & l_2 & l_3 & l_4 & l_5 & l_6 & l_7 \\ l_0 & l_1 & l_2 & l_3 & l_4 & l_5 & l_6 & l_7 \end{vmatrix}_{8 \times 8}$$

The abrupt change so formed in "z" can be illustrated as a 2D step function $\Delta(x, y)$, which is equal to

$$\Delta(x, y) = \begin{cases} -\dfrac{1}{M} & \text{for } x = 0, 1, \ldots M-1 \text{ and } y = 0, 1, \ldots \dfrac{M-2}{2} \\ \dfrac{1}{M} & \text{for } x = 0, 1, \ldots M-1 \text{ and } y = 4, 5, \ldots M-1 \end{cases} \quad (11.5)$$

Let $\mathbf{a}_v$ be the amplitude of $f(x, y)$ and δ_m be slope of $f(x, y)$, then block "z" can be modeled as

$$Z(x, y) = a_v + \delta_m \cdot b(x, y) + \beta \cdot \Delta(x, y) + r'(x, y) \\ \text{for } x, y = 0, 1, \ldots (N-1) \quad (11.6)$$

where a_v is the average value of "z" and it represents brightness of the local background, $\acute{r}(x, y)$ is the residual block, which represents the activity across block boundaries and edges. β represents the amplitude of step function $\Delta(x, y)$ and this represents the seriousness of blocking artifacts within the compressed images. The average value a_v can be calculated[7] as

$$a_v = \frac{\text{DC value of } Z}{N}$$

Let us consider that the difference between two horizontal pixels is represented by $d(x, y)$ such that

$$d(x, y) = Z(x, y) - Z(x, y-1) \quad (11.7)$$

From eq 11.6, the slope of the left half of "z" is represented by δ_{m_L} such that

Detection of Blocking Artifacts in JPEG Compressed Images

$$\delta_{m_L} = -\frac{1}{M}\sum_{x=0}^{M-1}\left[\frac{64}{M-4}\sum_{y=1}^{(M/2)-1} d(x, y)\right] \quad (11.8)$$

Consequently, the slope of "z" on right half is

$$\delta_{m_L} = -\frac{1}{M}\sum_{x=0}^{M-1}\left[\frac{64}{M-4}\sum_{y=1}^{(M/2)-1} d(x, y)\right] \quad (11.9)$$

The resultant slope can be obtained by averaging δ_{m_L} and δ_{m_R} by putting the values of δ_{m_L} and δ_{m_R} from eqs 11.8 and 11.9 as

$$\delta_m = \frac{\delta_{m_L} + \delta_{m_R}}{2} = \frac{24}{M}\sum_{x=0}^{M-1}\frac{\left[\sum_{y=1}^{M-1} d(x, y) - d(x, M/2)\right]}{M-4}$$

$$= 24\sum_{x=0}^{M-1}\frac{\left[\sum_{y=1}^{M-1} Z(x, y) - Z(x, y-1) - d(x, M/2)\right]}{M(M-4)} \quad (11.10)$$

Now if we calculate a_v and δ_m, remaining part of eq 11.6 can be calculated as $R(x, y) = \beta \cdot \Delta(x, y) + r'(x, y) + a_v$ and $\tilde{a}(x, y)$ that represent white Gaussian noise having mean energy value = 0. Thus, β can be calculated as

$$\beta = Z(x, y) - \Re(x, y) \quad (11.11)$$

Using eq 11.11, blocking artifacts can be measured quantitatively. Literature shows that blocking artifacts measuring algorithm cause a large computational problem while working in pixel-level operations. To avoid such issues, a simply and fast DCT algorithm is introduced, described in the next section.

11.3.3 FAST DCT ALGORITHM FOR BLOCKING ARTIFACTS

Let X, Y, and Z be the BDCT of "x," "y," and "z" blocks, respectively; let us consider two matrixes as follows:

$$m_1 = \begin{vmatrix} 0 & 0 & 0 & 0 & 0 & 0 & 0 & 0 \\ 0 & 0 & 0 & 0 & 0 & 0 & 0 & 0 \\ 0 & 0 & 0 & 0 & 0 & 0 & 0 & 0 \\ 0 & 0 & 0 & 0 & 0 & 0 & 0 & 0 \\ 1 & 1 & 1 & 1 & 0 & 0 & 0 & 0 \\ 1 & 1 & 1 & 1 & 0 & 0 & 0 & 0 \\ 1 & 1 & 1 & 1 & 0 & 0 & 0 & 0 \\ 1 & 1 & 1 & 1 & 0 & 0 & 0 & 0 \end{vmatrix}_{8 \times 8}$$

$$m_2 = \begin{vmatrix} 0 & 0 & 0 & 0 & 1 & 1 & 1 & 1 \\ 0 & 0 & 0 & 0 & 1 & 1 & 1 & 1 \\ 0 & 0 & 0 & 0 & 1 & 1 & 1 & 1 \\ 0 & 0 & 0 & 0 & 1 & 1 & 1 & 1 \\ 0 & 0 & 0 & 0 & 0 & 0 & 0 & 0 \\ 0 & 0 & 0 & 0 & 0 & 0 & 0 & 0 \\ 0 & 0 & 0 & 0 & 0 & 0 & 0 & 0 \\ 0 & 0 & 0 & 0 & 0 & 0 & 0 & 0 \end{vmatrix}_{8 \times 8} \quad (11.12)$$

Therefore,

$$\hat{z} = x \cdot m_1 + y \cdot m_2 \quad (11.13)$$

By applying BDCT, we will get

$$\widehat{Z} = X \cdot M_1 + Y \cdot M_2 \quad (11.14)$$

For simplification, let us assume $N = 8$. The BDCT transform of kth block of x_k of size 8×8 is defined as

$$X_k(u, v) = \Psi(u)\Psi(v) \sum_{x=0}^{N-1} \sum_{y=0}^{N-1} x_k(x, y) \otimes \Phi \otimes \Upsilon$$

where

$$\Phi = \cos\left(\frac{\pi(1+2x)u}{16}\right) \; \& \; \Upsilon = \cos\left(\frac{\pi(1+2x)v}{16}\right) \quad (11.15)$$

where $k = 1, 2$ and

$$\Psi(u) = \Psi(v) = \begin{cases} \dfrac{1}{\sqrt{N}} & \text{if } u = v = 0 \\ \dfrac{2}{\sqrt{N}} & \text{otherwise} \end{cases} \quad (11.16)$$

Let us consider that x_k contains variation in original pixel and it can be modeled by the slope (δ_m) of 2D linear function in both left and right sides of blocking edge. Now we have to measure the value of the slope by substituting the low-frequency component of x and y putting the value of slope in eq 11.15 for $u = 0$ and $v = 1$.

$$Z_k(0,1) = \frac{\sqrt{2}}{8} \sum_{x=0}^{7}[\pi + \Omega]$$

where

$$\pi = \sum_{y=0}^{(N/2)-1} \delta_k\left(\frac{N-24}{2} - \frac{y}{2}\right) \otimes \cos\left(\frac{\pi(1+2x)}{16}\right)$$

&

$$\Omega = \sum_{y=(N/2)}^{N-1} \delta_k\left(\frac{N-1}{2} - \frac{y-4}{2}\right) \otimes \cos\left(\frac{\pi(1+2x)}{16}\right) \quad (11.17)$$

$$= \eta \delta_m \quad (11.18)$$

where $\eta = -18.2241$ value calculated by Park et al.[11]. It is helpful in calculating the average value of slope as

$$\delta_m = \frac{\delta_{m_L} + \delta_{m_R}}{2} = \frac{X(0,1) + Y(0,1)}{2\eta} \quad (11.19)$$

From eq 11.19, one can easily calculate the average value of slope as two individual slopes are already available. To obtain the value of "β," let us represent the most important row of the 8 × 8 BDCT transform of $g(x, y)$ by $b = [l_0, l_1, \ldots, l_7]$. To calculate the value of "β," one has to calculate $\hat{B}(x, y)$ block as shown in the following equation

$$\hat{B}(x,y) = \begin{cases} B(x,y) - \delta_m \cdot k_y, & x = 0 \text{ and } y = 1, 2, \ldots 7 \\ 0 & x = y = 0, \\ B(x,y) & \text{elsewhere} \end{cases} \quad (11.20)$$

Note that the 8 × 8 DCT transform has only four nonzero components. Let the vector $m = [m_0, m_2, \ldots, m_7]$ be the first row of the 8 × 8 BDCT of the 2D step function, where $m_0 = m_2 = m_4 = m_6 = 0$. By the unitary property of the DCT, we have

$$|m| = \sqrt{\sum_{i=0}^{7} m_i^2} = \sqrt{\sum_{x=0}^{7}\sum_{y=0}^{7} \Delta^2(x, y)} = 1 \qquad (11.21)$$

So from eq 11.21, we can deduce the value of β as follows

$$\beta_{\text{Vertical}} = \sum_{k=0}^{7} m_k \hat{B}(0, k) \qquad \text{for } k = 1, 3, 5, 7 \qquad (11.22)$$

where β_{Vertical} blocking artifacts are across vertical boundaries. Horizontal artifacts can be calculated in the same fashion.

11.4 RESULTS AND DISCUSSION

The proposed method was applied to various images such as Peppers, Lena, Pentagon, and Bridge. All the four images were compressed using JPEG compression, and results were compared at various bit rates. Blocking artifacts are determined in vertical, horizontal, and average mode. The estimation of genuine blocking artifacts as computed by finding the distinction between unique pixel esteems over the square limit as artifacts are seen over the boundaries more overwhelmingly.

Table 11.1 demonstrates the outcomes for picture Pepper that the horizontal blocking artifacts ßH of the proposed technique are near the first blocking artifacts when contrasted with Park et al.,[11] Singh et al.,[12] and Singh et al.[13] The estimation of vertical blocking artifacts ßv and average artifacts ßAvg by the proposed is close to original values as compared to other methods.

Figure 11.4(a–c) shows the graphical representation of Table 11.1 in Matlab for image Pepper. The BPP are varied from 0.65 to 0.13 and the figures clearly show that the proposed method outperforms w.r.t. methods of Park et al.,[11] Singh et al.,[12] and Singh et al.[13]

Table 11.2 shows the results for image Lena that the blocking artifacts along horizontal ßH, as well as vertical ßv and average ßAvg of the proposed method, are very close to the true value blocking artifacts as compared to methods of Park et al.,[11] Singh et al.,[12] and Singh et al.[13]

TABLE 11.1 Comparison of Blocking Artifacts of Proposed Method with Others for Image Pepper.

Pepper image	Original			Park et al. [11]			Singh et al. [12]			Singh et al. [13]			Proposed		
BPP	βH	βv	βAvg	βH	βv	βAvg	βH	βv	βAvg	βH	βv	βAvg	βH	βv	βAvg
0.65	42.96	35.23	38.53	34.87	27.34	32.9	43.21	33.24	37.90	44.67	34.56	38.23	40.06	33.08	36.57
0.55	43.12	35.46	38.69	35.23	27.65	33.20	44.45	34.76	37.80	45.67	35.56	38.70	41.40	34.63	38.02
0.45	43.56	38.78	39.78	35.78	29.36	33.98	45.87	35.76	39.67	46.78	36.78	40.34	44.76	37.87	42.50
0.36	47.87	41.98	43.67	38.45	34.45	37.76	46.54	37.67	39.87	47.23	38.76	40.34	51.11	43.01	47.06
0.26	64.78	51.35	52.34	43.34	36.45	38.78	58.87	38.78	43.46	59.78	39.98	44.78	67.17	55.89	51.53
0.15	97.67	64.35	69.87	51.34	41.24	43.67	59.97	46.78	49.67	57.65	45.68	46.45	72.95	58.04	65.49
0.13	98.34	65.23	70.54	52.23	43.23	45.78	61.45	37.45	50.12	59.45	47.98	47.67	74.14	62.55	68.34

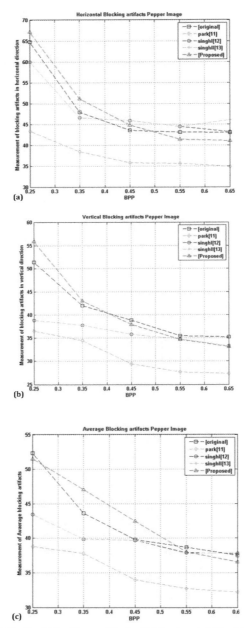

FIGURE 11.4 (a) Vertical blocking artifacts, (b) horizontal blocking artifacts, and (c) average blocking artifacts measured for Pepper image.

Detection of Blocking Artifacts in JPEG Compressed Images

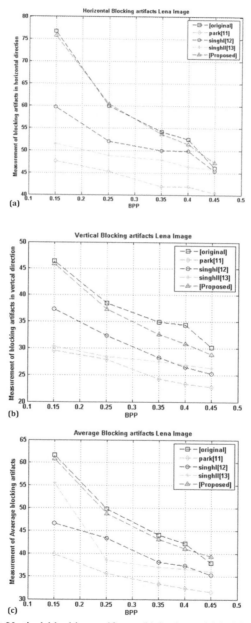

FIGURE 11.5 (a) Vertical blocking artifacts, (b) horizontal blocking artifacts, and (c) average blocking artifacts measured for Lena.

TABLE 11.2 Comparison of Blocking Artifacts of Proposed Method with Others for Image Lena.

Lena image BPP	Original			Park et al. [11]			Singh et al. [12]			Singh et al. [13]			Proposed		
	ßH	ßv	ßAvg	ßH	ßv	ßAvg	ßH	ßv	ßAvg	ßH	ßv	ßAvg	ßH	ßv	ßAvg
0.46	46.11	30.10	37.97	40.41	22.78	31.59	45.39	25.23	35.31	45.79	26.25	36.02	47.32	28.83	39.33
0.40	52.45	34.34	42.23	41.89	23.37	32.45	49.88	26.48	37.28	46.39	26.88	36.63	51.39	30.82	41.10
0.35	54.18	34.88	44.03	41.96	24.25	33.34	49.95	28.20	39.19	47.90	27.54	36.98	53.68	32.59	43.13
0.25	60.00	38.35	49.76	45.35	27.85	35.54	52.03	32.35	43.36	48.78	28.36	38.57	60.43	37.24	48.83
0.14	76.77	46.34	61.68	47.67	29.56	39.87	59.65	37.27	46.67	51.55	30.55	41.05	75.77	45.87	60.82

Figure 11.5(a–c) shows the graphical representation of Table 11.2 in Matlab for image Lena. The BPP are varied from 0.46 to 0.14 and the figures clearly show that the proposed method outperforms w.r.t. methods of Park et al.[11], Singh et al.,[12] and Singh et al.[13]

Table 11.3 shows the results for image Pentagon that the blocking artifacts along horizontal ßH, as well as vertical ßv and average ßAvg of the proposed method, are very close to the true value blocking artifacts as compared to methods of Park et al.[11], Singh et al.,[12] and Singh et al.[13]

Figure 11.6(a–c) shows the graphical representation of Table 11.3 in Matlab for image Pentagon. The BPP are varied from 0.85 to 0.21 and the figures clearly show that the proposed method outperforms w.r.t. methods of Park et al.[11], Singh et al.,[12] and Singh et al.[13]

Table 11.4 shows the results for image Bridge that the horizontal blocking artifacts ßH of the proposed method are very close to the original blocking artifacts as compared to Park et al.[11], Singh et al.,[12] and Singh et al.[13] The values of vertical blocking artifacts ßv and average artifacts ßAvg by the proposed are close to original values as compared to methods of Park et al.[11], Singh et al.,[12] and Singh et al.[13]

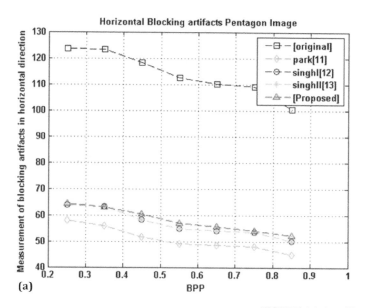

FIGURE 11.6 *(Continued)*

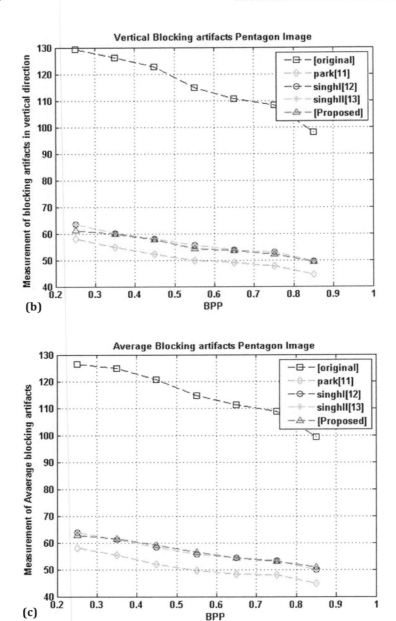

FIGURE 11.6 (a) Vertical blocking artifacts, (b) horizontal blocking artifacts, and (c) average blocking artifacts measured for Pentagon.

TABLE 11.3 Comparison of Blocking Artifacts of Proposed Method with Others for Image Pentagon.

Pentagon image	Original			Park et al. [11]			Singh et al. [12]			Singh et al. [13]			Proposed		
BPP	βH	βv	βAvg	βH	βv	βAvg	βH	βv	βAvg	βH	βv	βAvg	βH	βv	βAvg
0.85	100.4	97.94	99.30	44.95	44.61	44.78	50.08	49.48	49.78	50.08	49.48	49.78	52.35	49.32	50.84
0.75	109.1	108.3	108.7	48.10	47.85	47.97	53.57	52.97	53.27	53.57	52.97	53.27	54.05	52.13	53.09
0.66	110.2	110.6	111.3	48.50	48.97	48.23	54.20	53.80	54.30	54.20	53.80	54.30	55.54	53.45	54.34
0.55	112.5	114.8	114.6	49.20	49.87	49.67	54.90	55.76	55.70	54.90	55.76	55.70	56.89	54.34	56.50
0.43	118.4	122.6	120.5	51.78	52.35	52.07	58.34	58.03	58.18	58.34	58.03	58.18	60.44	57.65	59.04
0.36	123.3	126.2	124.7	55.96	54.80	55.38	63.36	60.21	61.28	63.36	60.21	61.28	63.08	59.96	61.52
0.25	123.5	129.3	126.3	57.93	58.13	58.03	63.90	63.62	63.76	63.90	63.62	63.76	64.30	61.26	62.78
0.21	127.8	130.0	128.9	59.10	58.90	59.00	64.70	64.30	64.50	64.70	64.30	64.50	66.25	62.53	64.39

TABLE 11.4 Comparison of Blocking Artifacts of Proposed Method with Others for Image Bridge.

Bridge image	Original			Park et al. [11]			Singh et al. [12]			Singh et al. [13]			Proposed		
BPP	ßH	ßv	ßAvg	ßH	ßv	ßAvg	ßH	ßv	ßAvg	ßH	ßv	ßAvg	ßH	ßv	ßAvg
0.86	67.65	75.57	73.30	58.34	71.78	63.36	63.35	81.52	71.75	63.35	81.52	71.75	63.55	81.49	72.52
0.78	70.67	75.87	73.56	58.65	70.35	64.14	64.40	81.96	72.34	64.40	81.96	72.34	64.97	82.95	73.96
0.66	76.54	82.45	73.85	59.45	73.35	64.67	65.25	82.13	73.34	65.25	82.13	73.34	65.32	84.69	75.01
0.55	78.56	87.68	74.95	59.89	73.70	65.14	65.78	82.78	73.88	65.78	82.78	73.88	66.67	85.12	75.89
0.43	81.01	91.20	88.80	60.69	74.45	65.45	66.95	83.23	74.67	66.95	83.23	74.67	70.56	88.48	79.52
0.36	89.67	101.3	91.23	61.12	78.58	69.97	70.86	88.56	78.88	70.86	88.56	78.88	71.74	89.91	80.83
0.28	101.2	112.3	108.5	61.34	79.67	70.45	71.95	88.78	79.89	71.95	88.78	79.89	71.84	90.75	80.95

Detection of Blocking Artifacts in JPEG Compressed Images

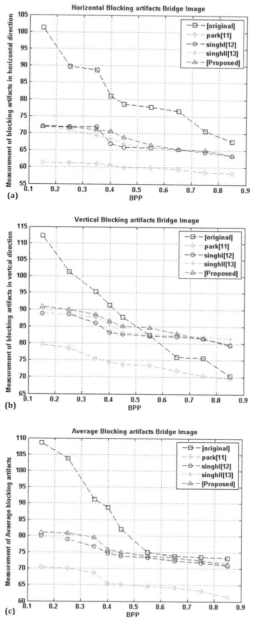

FIGURE 11.7 (a) Vertical blocking artifacts, (b) horizontal blocking artifacts, and (c) average blocking artifacts measured for Bridge.

Figure 11.7(a–c) shows the graphical representation of Table 11.4 in Matlab for image Bridge. The BPP are varied from 0.85 to 0.25 and the figures clearly show that the proposed method outperforms w.r.t. methods of Park et al.[11], Singh et al.,[12] and Singh et al.[13]

To determine the precision of the proposed method, %age absolute error acts as a useful tool and is calculated by

$$\%\text{age error} = \left|\frac{\#\text{experimental} - \#\text{theoratical}}{\#\text{experimental}}\right| \times 100 \quad (11.23)$$

Table 11.5 depicted that %age error of proposed method is very close to zero or relatively close to zero w.r.t. other methods used in literature. Figure 11.8(a–d) shows graphical representation of %age absolute error.

TABLE 11.5 Results of Percentage Absolute Error for Different Images (512 × 512) w.r.t. Original Value of Beta Average.

Image	BPP	Park et al. [11]	Singh et al. [12]	Singh et al. [13]	Proposed
Pepper	0.65	14.61	1.64	1.64	1.51
	0.55	14.19	2.30	2.30	1.74
	0.36	13.53	8.70	8.70	7.77
	0.26	25.91	16.97	16.97	1.55
Lena	0.46	16.80	7.01	5.14	3.57
	0.40	23.16	11.73	13.26	2.67
	0.35	24.27	10.99	16.00	2.03
	0.25	28.58	12.86	22.49	1.86
	0.14	35.36	24.34	33.45	1.39
Pentagon	0.85	54.90	49.87	49.87	48.81
	0.75	55.87	50.99	50.99	51.16
	0.43	56.82	51.75	51.75	51.03
	0.36	55.62	50.89	50.89	50.70
Bridge	0.86	13.56	2.11	2.11	1.06
	0.78	12.81	1.66	1.66	0.54
	0.66	12.43	0.69	0.69	1.57
	0.55	13.09	1.43	1.43	1.26
	0.43	26.30	15.91	15.91	10.45
	0.36	23.30	13.54	13.54	11.40

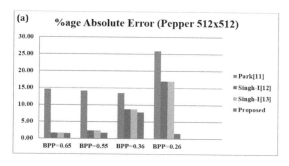
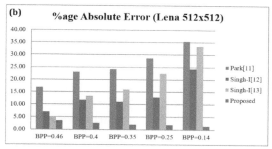
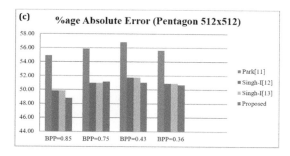
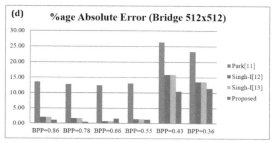

FIGURE 11.8 %age absolute error at different BPP for (a) Pepper, (b) Lena, (c) Pentagon, and (d) Bridge.

11.5 CONCLUSION

A block-based DCT domain blind measurement method for detection of blocking artifact is discussed and is applied to the variety of images. The function so designed gives better performance than the predefined methods of Park et al.[11], Singh et al.,[12] and Singh et al.[13] using different indices like ßH, ßv, and ßAvg as shown in Tables 11.1–11.3. The proposed method was implemented on various test images with the help of 2D function at various BPP using JPEG compression, and computational results clearly show that the designed 2D function is better than the earlier detection methods of Park et al.[11], Singh et al.,[12] and Singh et al.[13], as it gives close proximity results than other methods. The validation of the proposed method is further done by calculating percentage absolute error as shown in Table 11.5. In the case of images with fewer details like Pepper, Lena, and Bridge, the detection process of the proposed method is approximately greater or equal to 90% efficient as compared to other methods that give detection of around 70% clearly shows. The proposed method works fairly well in the case of images with more details like Pentagon w.r.t. to other postprocessing methods. The proposed method can be used for real-time image/videos due to its low computational requirements and hence can be used for multimedia and IoT applications.

DISCLOSURES

The authors declare that there is no conflict of interest regarding the publication of this chapter.

KEYWORDS

- image compression
- blocking artifacts
- DCT
- BPP
- % age absolute error

REFERENCES

1. ITU Recommendations H.261. Video Code for Audio Visual Service at p x 64 k bits/sec; 1993.
2. Wijewardhana, U.; Codreanu, M. In *Lapped Transforms Based Image Recovery for Block Compressed Sensing*, IEEE Conference on Data Compression, March 2018.
3. Bhardwaj, D.; Pankajakshan, V. A JPEG Blocking Artifact Detector for Image Forensics. *Signal Process. Image Commun.* **2018**, *68*, 155–161.
4. Dalmia, N.; Okade, M. Robust First Quantization Matrix Estimation Based on Filtering of Recompression Artifacts for Non-aligned Double Compressed JPEG Images. *Signal Process. Image Commun.* **2018**, *61*, 9–20.
5. Meier, T.; Ngan, K. N.; Cheng, G. Reduction of Blocking Artifacts in Image and Video Coding. *IEEE Trans. Circuits Syst. Video Technol.* **1999**, *9*, 490–500.
6. Brahimi, T.; Laouir, F.; Boubchir, L. et al. An Improved Wavelet-based Image Coder for Embedded Greyscale and Colour Image Compression. *Int. J. Electron. Commun.* **2017**, *73*, 183–192.
7. Wang, Z.; Bovik, A. C. In *Blind Measurement of Blocking Artifacts in Images*, Proceedings of the IEEE International Conference of Image Processing, Vancouver, Canada, Oct 2002; pp 981–984.
8. Liu, S. Z.; Bovik, A. C. Efficient DCT-Domain Blind Measurement and Reduction of Blocking Artifacts. *IEEE Trans. Circuits Syst. Video Technol.* **2002**, *12* (12), 1139–1149.
9. Luo, Y.; Ward, R. K. Removing the Blocking Artifact of Block-based DCT Compressed Images. *IEEE Trans. Image Process.* **2003**, *12* (7), 838–842.
10. Wang, C.; Zhang, W. J.; Fang, X. Z. Adaptive Reduction of Blocking Artifacts in DCT Domain for Highly Compressed Images. *IEEE Trans. Consum. Electron.* **2004**, *50*, 647–654.
11. Park, C. S.; Kim, J. H.; Ko, S. J. Fast Blind Measurement of Blocking Artifacts in Both Pixel and DCT Domains. *J. Math. Imaging Vis.* **2007**, *28*, 279–284.
12. Singh, J.; Singh, S.; Singh, D.; Uddin, M. Detection Method and Filters for Blocking Effect Reduction of Highly Compressed Image. *Signal Process. Image Commun.* **2011**, *26*, 493–506.
13. Singh, J.; Singh, S.; Singh, D.; Uddin, M. Blocking Artefact Detection in Block-based DCT Compressed Images. *Int. J. Signal Imaging Syst. Eng.* **2011**, *4* (3), 181–188.
14. Singh, J.; Singh, D.; Uddin, M. Detection Methods for Blocking Artefacts in Transform Coded Images. *IET Image Process.* **2014**, *8* (8), 435–444.

CHAPTER 12

Quality of Service Provisioning in Mobile Ad Hoc Networks

MANWINDER SINGH* and KAMAL KUMAR SHARMA

*School of Electronics and Electrical Engineering,
Lovely Professional University, Punjab, India*

*Corresponding author. E-mail: manwinder.25231@lpu.co.in

ABSTRACT

Delay-tolerant network (DTNs) are divided mobile ad hoc networks with discontinuous availability. DTNs are network conditions that are liable to postponements and disturbances. The DTN design targets giving executions to dependable message conveyance in discontinuously associated networks. The DTN design additionally determines the pack convention which controls the trading of packs. In this proposed work, an approach has been suggested that will adequately move the information with decreased loss of bundles. The proposed approach for directing builds the energy effectiveness of the organization. The result showed that the improvement in different execution boundaries in the organization. The experimental result demonstrates that the proposed approach is proficient as analyzed to the current methodology.

12.1 INTRODUCTION

A mobile ad hoc network (MANET) is a network in a mobile environment enabling devices to create and join networks on the fly. MANETs are also called mobile, to connect with various networks, they use wireless connections. It may be a normal Wi-Fi connection, or any other channels like cellular or satellite transmission system. In automobile sector, traffic congestion can be calculated from tracking of trucking fleets. Due to MANETs are dynamic in nature, they are normally not very reliable. So, it is very essential to be attentive to what information is forwarded over a MANET.[1-5] In multihop mobile networks, MANET nodes provide end-to-end connectivity throughout the networks. MANETs have some important attributes such as limited bandwidth and energy, confined (restricted) physical reliability, and dynamic topologies and because of these, routing protocols designed for wired networks cannot be straightly implemented for wireless networks also. MANET protocol substitutes routing message among nodes and sustains the routing states accordingly. Sivakumar et al.[6] proposed delay-tolerant network (DTN) protocol for the purpose of evaluation and testing. This proposed DTN protocol relied on the network virtualization. Additionally, in this paper, TUNIE architecture was also presented, which was competent for simulating consistent DTN environments and attaining a precise system performance evaluation. Lin and Liu[7] proposed a reliable protocol in DTN based on the human mobility model. This paper discussed the performances of well-known PROPHET and Spray and Wait routing protocol. In order to simulate the protocol, MATLAB simulator was used. The results obtained from simulation depicted that the proposed protocol outperforms w.r.t. other methods in terms of average delay, delivery ratio, and communication overhead. Lin[8] proposed delay-tolerant arrangement that was utilized to illustrate the already existing mechanism in congestion control. The proposed mechanism provided assistance to locate the DTN congestion control design space. On that paper, comparison had been done between the existing mechanism and the proposed architecture. Investigation of design space helped to recognize significant problems and issues that were required to be addressed. Liao et al.[9] presented a DTN for censorship-resistant communication. They analyzed the performance of the given system while developing an energy-efficient system. In their work, a flooding protocol was introduced while adoption rate was implemented. The simulation results showed that delay and its delivery rate were vigorous to denial of service. Gerasimov and Simon[10] presented routing protocol in DTN. In their work, dynamic quota mechanism had been proposed, which permitted routing to operate efficiently with distinct traffic loads. Those networks that were subjected to disruption and delays were basically DTNs. Due to long delay and intermittent connection, these end-to-end routing protocols are failure in these network circumstances. The simulation result indicates that per-hop transferring of many mimeos of the similar information to the destination may create suitable routing performance in DTNs. In their research work, proposed method was based on the setting of quota value to bind the number of copies. On that paper, probability-based system had also been proposed, which was low in cost. This probability-based method was used to eliminate the irrelevant or pointless message copies from the network. Dharmaraju et al.[11] proposed a fuzzy logic–based transport protocol to control network congestion in terms of traffic and reduced the frame quality to an acceptable level. Congestion factor was calculated using fuzzy logic on the basis of frame size and buffer size. Frame rate–based congestion controller had been used to maintain the rate of flow. Chen and Heinzelman[12] gave a cross-layer dynamic adaptation mechanism for wireless ad hoc networks. A joint congestion control scheme with scheduling algorithm was proposed for dynamic networks by changing scheduling scheme with adaptation model. Results obtained during simulation show that the proposed protocol is highly stable and robust for unicast data and it improves the packet relay in all mobility situations. Singh and Raghavendra[13] provided a congestion control mechanism by introducing a mobile sink to avoid congestion and improve network lifetime. They analyzed the consequence of mobile sink in decreasing congestion and increasing lifetime of the sensor network and effectively achieved better lifetime while avoiding congestion of wireless sensor network (WSN). Toh[14] proposed a swarm intelligence–based network that was highly efficient during traffic congestion specifically for WSNs by mimicking the obstacle avoidance behavior of birds flock. Evaluation outcomes have shown that the current approach performance is better in load balancing, powerfulness against defected nodes, and scalability. Chen and Nahrstedt[15] proposed Redundancy Aware Hierarchical Tree Alternative Path algorithm for congestion control in WSNs.

12.2 PROPOSED METHOD

In the proposed method, Artificial Bee Colony (ABC) algorithm incorporating Optimized Link State Routing (OLSR) protocol has been introduced. The given system is developed to minimize routing-related issues and make system more energy efficient. The basic steps followed to develop system in simulated environment specifically for MANET are explained later. These steps are implemented and shown in flowchart of methodology in Figure 12.1.

Step 1: To design a MANET network.

Step 2: Initialize n nodes within the network.

Step 3: Describe source node and sink node from the set of all the available nodes of MANET.

Step 4: Initialize a number of rounds while defining coverage area of each node within MANET.

Step 5: Using OLSR protocol to define a specific source to sink route.

Step6: As per the requirement of MANET, initialize ABC algorithm with an objective/fitness function such that

$$Fit_{Function} = \begin{vmatrix} F_s & \text{if } F_s \geq F_t \\ F_t & \text{else} \end{vmatrix} \quad (12.1)$$

where F_s is the selected properties of the individual nodes and the F_t is the threshold properties and it is the average of the properties of all nodes.

Step 7: To optimize the best route from source to sink for successful packet delivery, ABC algorithm is used.

Step 8: If for a particular node, fitness functions fail, apply artificial neural network (ANN) and set the rules to identify suspected node. Discard the malicious node and start data transmission through other nodes.

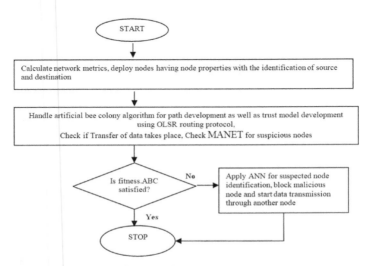

FIGURE 12.1 Methodology of proposed work.

12.3 SIMULATION RESULTS

To validate the proposed algorithm, we have developed a MANET network with 50 nodes distributed randomly over a particular area. Width and height of the network area is considered as 1000. The number of rounds = 5 is considered to calculate various parameters for evaluation.

Figure 12.2 depicts the energy consumption parameters of proposed technique w.r.t. conventional method (hotspot-based approach) and the comparison was taken against time interval. The proposed technique efficiently detects suspected nodes and reduces packet retransmission rate, thereby reducing energy consumption as compared to iHAR (improved Hotspot-based Adaptive Routing) method.

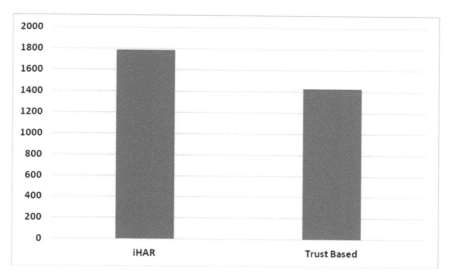

FIGURE 12.2 Energy consumed in the network.

In Figure 12.3, packet delivery ratio (PDR) of proposed method is calculated w.r.t. time, and the results are compared with the state-of-the-art method. In these results, a number of nodes are kept constant, and its value is fixed at 29. Results in Figure 12.3 clearly show the creativeness of the proposed method w.r.t. the iHAR state-of-the-art method.

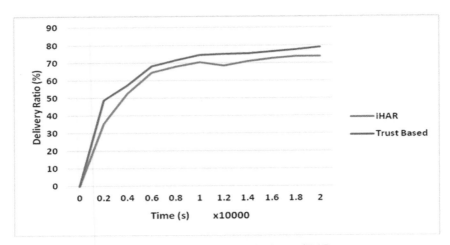

FIGURE 12.3 PDR versus time of proposed method w.r.t. iHAR.

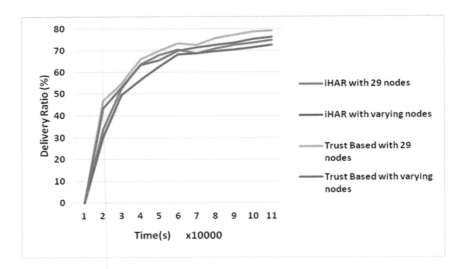

FIGURE 12.4 Packet delivery ratio (with flexible amount of nodes).

Figure 12.4 illustrates PDR w.r.t. time with variable number of nodes using iHAR and proposed method. On the other hand, Figure 12.5 depicts delay function for different nodes for proposed method w.r.t. the state-of-the-art (iHAR) method. Figures 12.4 and 12.5 clearly show trust-based

Quality of Service Provisioning in Mobile Ad Hoc Networks 203

proposed technique that outperforms existing methods with fixed as well as variable number of nodes.

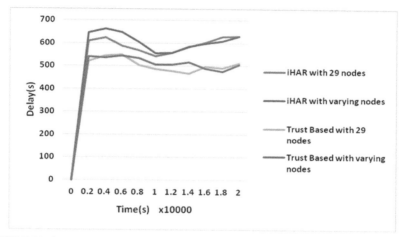

FIGURE 12.5 Packet delivery delay (with different node numbers).

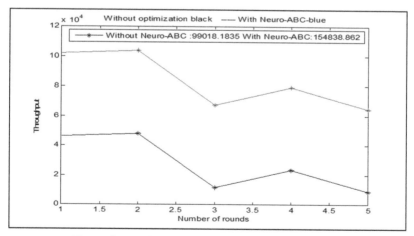

FIGURE 12.6 Throughput without optimization and with neuro-ABC.

Figure 12.6 illustrates throughput of proposed method (with ANN–ABC) w.r.t. a number of rounds for the MANET system. From the figure, it is very much clear that we have achieved approximately 3 times throughput w.r.t. the traditional system that can achieve average throughput up to 3.1×10^4 for five rounds of simulation.

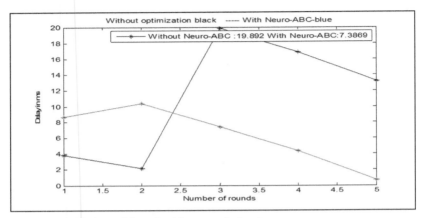

FIGURE 12.7 Delay with and without optimization.

Figure 12.7 depicts delay (ms) of proposed method (with ANN–ABC) w.r.t. a number of rounds for the MANET system. From the figure, it is very much clear that we have achieved approximately 1/2 time in terms of delay (6.42 ms) w.r.t. the traditional system that can achieve average delay of 11.24 ms.

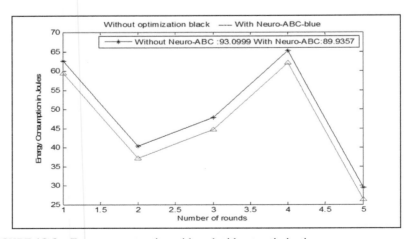

FIGURE 12.8 Energy consumption with and without optimization.

While sending the signal from one to another, every node would take some of the energy. The static node takes less amount of energy, whereas busy node takes more amount of energy. Therefore, it is significant to find the route with less power consumption for data transmission. So, the routing protocol that considers residual energy improves performance in the contrast to protocols that do not. The results are executed when the protocol is occupied with the network and when the classifiers or the optimization are considered in the network as depicted. It is being concluded that more energy consumption has been obtained using ABC algorithm than without optimization when ABC has not been utilized. Figure 12.8 illustrates energy consumed by the proposed method approximately 10% less as compared to the state-of-the-art method. The numerically proposed system utilizes 45.8 J of energy w.r.t. 50 J, which is used during the traditional approach.

12.4 CONCLUSION

A novel approach for the computation of these parameters is used. The approach is based on the opinions of the other nodes in the network. These opinions are the values of parameters like number of packets lost by the target node or number of transmissions and the trust value. The intermediary node selection is based on the computation approach. The results of the proposed method are compared with traditional method (iHAR) using different performance parameters such as PDR, packet delivery delay, and energy consumption. The comparison shows that the proposed approach outperforms the hotspot-based approach by a significant margin. The result also validates the proposed technique and shows the correctives and advantages like less energy consumption, more PDR with less delay as compared to the state-of-the-art method.

The most important thing in finding the route in the DTNs is the energy efficiency of the approach. The energy efficiency must be taken into account every time an algorithm is selected for the calculations. The less the calculation, the more efficient approach must be. The number of computations on every node must also increase the end-to-end delay of the packets in the network, which is completely undesirable for an approach that works on straightforward node data selection and forwarding. The approach must also consider the fact that the data to be transferred must be filtered to avoid unwanted messages being transferred and utilizing unnecessary bandwidth of the network. Bloom filter is one of the examples of such filters. Besides that, other machine learning algorithms must be used for the calculation of fairness of the node to which data to be transferred.

The simulations of the proposed mechanism run on MATLAB with 50–100 nodes. The average throughput with and without optimization was obtained. With optimization, the 61% of delay is decreased, which is better for a network. The energy consumed by the network with optimization was less than the without optimization network. It was reduced up to 8.4%.

In future, different routing protocols like Ad hoc On-Demand Distance Vector, Dynamic Source Routing, and Temporally Ordered Routing Algorithm and hybrid routing protocols like ZRP, BGP, and EIGRP can also be used. For optimization, different algorithms such as genetic algorithm and particle swarm optimization will be used. For classification, algorithms like fuzzy logic and support vector machine will be used.

KEYWORDS

- **MANET**
- **EED**
- **PDR**
- **energy consumption**

REFERENCES

1. Mohapatra, P.; Jian, L.; Gui, C. QoS in Mobile Ad Hoc Networks. *IEEE Wireless Commun. Mag.* **2003**, *10* (3), 44–52.
2. Karn, P. In *MACA—A New Channel Access Method for Packet Radio*, ARRL/CRRL Amateur Radio 9th Computer Networking Conference, April 1990; pp 134–140.
3. Bharghavan, V.; Demers, A.; Shenker, S.; Zhang, L. MACAW: A Media Access Protocol for Wireless LAN's. *ACM SIGCOMM Comput. Commun. Rev.* **1994**, *24* (4), 212–225.
4. Chakrabarthi, S.; Mishra, A. QoS Issues in Ad Hoc Wireless Networks. *IEEE Commun. Mag.* **2001**, *39* (2), 142–148.
5. Zhang, X.; Le, S. B.; Gahng, A.; Campbell, A. T. INSIGNIA: An IP Based Quality of Service Framework for Mobile Ad Hoc Networks. *J. Parallel Distrib. Comput.* **2000**, *60* (4), 374–406.

6. Sivakumar, R.; Sinha, P.; Bharghavan, V. CEDAR: A Core Extraction Distributed Ad Hoc Routing Algorithm. *IEEE J. Sel. Areas Commun.* 1999, *17* (8), 1454–1465.
7. Lin, C. R.; Liu, J. QoS Routing in Ad Hoc Wireless Networks. *IEEE J. Sel. Areas Commun.* **1999,** *17* (8), 1426–1438.
8. Lin, C. R. On-Demand QoS Routing in Multihop Mobile Networks. *Proc. IEEE Int. Conf. Comput. Commun. (INFOCOM)* **2001,** *3*, 1735–1744.
9. Liao, W. H.; Tseng, Y. C.; Wang, S. L.; Sheu, J. P. A Multi-Path QoS Routing Protocol in a Wireless Mobile Ad Hoc Network. *Proc. IEEE Int. Conf. Netw.* **2001,** *2*, 158–167.
10. Gerasimov, I.; Simon, R. In *A Bandwidth Reservation Mechanism for On-Demand Ad Hoc Path Finding*, Proceedings of the IEEE/SCS 35th Annual Simulation Symposium, April 2002; pp 27–34.
11. Dharmaraju, D.; Roy Chowdhury, A.; Hovareshti, P.; Baras, J. S. In *INORA—A Unified Signaling and Routing Mechanism for QoS Support in Mobile Ad Hoc Networks*, Proceedings of the 31st International Conference on Parallel Processing Workshops, August 2002; pp 86–93.
12. Chen, L.; Heinzelman, W. B. QoS-Aware Routing Based on Bandwidth Estimation for Mobile Ad Hoc Networks. *IEEE J. Sel. Areas Commun.* **2005,** *23* (3), 561–572.
13. Singh, S.; Raghavendra, C. S. PAMAS—Power Aware Multi Access Protocol with Signaling for Ad Hoc Networks. *ACM Comput. Commun. Rev.* **1998,** *28* (3), 5–26.
14. Toh, C. K. Maximum Battery Life Routing to Support Ubiquitous Mobile Computing in Wireless Ad Hoc Networks. *IEEE Commun. Mag.* **2001,** *39* (6), 138–147.
15. Chen, S.; Nahrstedt, K. Distributed Quality-of-Service Routing in Ad Hoc Networks. *IEEE J. Sel. Areas Commun.* **1999,** *17* (8), 1488–1505.

CHAPTER 13

Solution of Automatic Generation Control of Multi-Area Power Plant Strategy with a Nonconventional Energy Source in Cooperation with Smart Controllers

KRISHAN ARORA* and TARUN DHANDHEL

*School of Electronics and Electrical Engineering,
Lovely Professional University, Punjab, India*

*Corresponding author. E-mail: Krishan.12252@lpu.co.in

ABSTRACT

The exhaustible energy sources that are limited and generate themselves in nature are known as nonconventional energy or renewable energy source. It includes tidal energy, solar energy, and wind energy. Fossil fuels that can be named conventional energy sources are limited stock in nature; with the continuous use of fossil fuels, it would go extinct. Since the evolution and improvement of mankind are very close to energy sources, most of the countries throughout the world have enlisted in pursuit and exploration of nonconventional energy sources that are very useful to sustain the human life cycle. The major command within a utility's energy control center is automatic generation control (AGC), the aim of which is to keep tracking the load variations with retaining the system frequency, tie-line interchanges, and best generation level close to specified values so that nominal frequency and tie-line schedules are sustained. This can be accomplished with practical and smart controllers. The AGC is amalgamation of wind, hydro, and thermal power plants, which has connected with progressive controllers. Particle swarm optimization is highly productive and reliable in calculations of dissimilar gains in load-frequency control. Moreover, particle swarm optimization recital is much better than gradient descent to get the best-optimized result for the discrete parameter and controller, which enhanced vital performance for three area networks.

13.1 INTRODUCTION

The demand of continual energy sources, wind, hydro, and solar, is increasing day by day with increasing technology. There are main merits of the conventional energy sources that are summarized as:

- Fossil resources are dependent, decreasing with time.
- Environmental protection and greenhouse gas emissions are reduced.
- Energy production cost is reduced.

Therefore, the problems of load-frequency control (LFC) need to be considered as important in power plants. With the huge availability of the wind power in the nature, the LFC problems are difficult to incorporate in wind power plant. This issue can be sorted out with the investigation of prediction of wind direction and speed to get the idea with the available wind power. The conventional energy sources are not fixed and not easily available in the nature, and the independent energy sources with overestimation and underestimation must be available in wind power plant as a function of cost. Particle swarm optimization (PSO) approach is tested on economic dispatch obstacle with a complicated amount of operations.

Here new ideology of LFC problems, including wind power plant with the help of PSO method, is presented. Wind, hydro, and thermal power plants mainly depend on natural resources. Furthermore, wind energy plays an indispensable role in the wind power plant, which also depends on the tariff of the power plant. So, overall expenditure can be determined with the help of wind power at the particular instant of time. Therefore, mathematical problems cannot tell about the real optimum economic dispatch. PSO method is made to corporate with wind power plants' economic. The combination of the power plants depicts the total cost differently simulated scheme with or without the production of power.

Operating frequency, load flow, and voltage conditions must maintain constant throughout electrical power transfer in electrical power plant. With the help of the abovementioned parameters, electrical power can be preserved constantly during the operating hour. There must be a perfect balance in power requirement and power generation to grant the full capacity power to the customers with affordable, finest electric power. Genetic algorithm (GA) is one of the approaches used to optimize controller gains other than artificial neural network, particle swarm technique, and fuzzy logic.[1] The variation in the tractable sources in system and generation of actual and imaginary power can be controlled properly. The smallest variation in the load side changes the operating output of the power system. The variation in power interest mainly affects frequency of power system and power flows in tie-line in control area. In power system, reliability criteria depend consistently on generation and power demand.[2]

The main aim is to keep the frequency constant all the time by varying generator units with the help of area control error (ACE), it can be controlled to zero by making the continual alteration of operating power.[3] ACE is amalgamation of the frequency and frequency time aberration and power networking error that can be used in an input of load-frequency controller; this method is used to regulate area to control error and variation in linkage frequency.

LFC methods are used in industry where proportional integral (PI) type and online base approach are used in which trial and error is the main objective to know about the problems.[4] To optimize control in different parameters, there are various optimization techniques that are probing in real-time simulation not for a single but for whole electrical power system. Two nonidentical power plants with two interconnected area power systems are regarded for research. The hydro and nuclear power plants can be considered as field-1 and field-2, respectively. With the help of the control strategies mentioned earlier for testing the controller in nonreal-time situation, it can be converted into real-time controller, which can be used to get the real-time results. The real-time simulation gives the ideology of producing a prototype into real function for industrial processes and devices that can be managed with an intertwined control system.[5] Day by day, load-frequency controllers are made with less tendency. The controller is having the advanced technology of strong control strategy known as PSO to get the best possible result with expected output. Gradient descent is also one of the methods that get tuned with various parameters. The best way to select governor speed supervision criterion is to calculate zero steady-state error present in frequency. The variation in the speed governor also helps to get the best-optimized result in automatic generation control (AGC). In PSO, firstly, the designed controller works according to the fuzzy logic rules such that it works fundamentally strong. Secondly, earlier fuzzy logic controller (FLC) was optimized with PSO so as to get the best possible result with adjustment of membership functions.[6]

13.1.1 NUTSHELL OF AGC PLOT

The flywheel and governor are the main part of the synchronous machine that makes it possible to achieve the desired power. This concept was not as sufficient as one can rely on. So, that governor needs to be controlled with a second method that is proportional to the variation in frequency in integral. There are lots of articles presented on AGC that helps to find out the main idea and creates a development in this area to control the transmission and distribution (Fig. 13.1).[7]

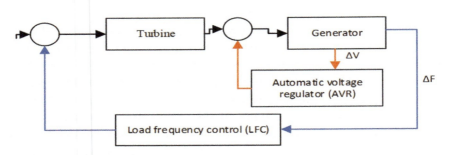

FIGURE 13.1 Automatic generation control.

The ACE can be controlled with the help of making a secondary AGC scheme. There are many aspects that can be seen in the chapter regarding the power system control models, AGC infrastructure, and a few numbers of controlling approaches that are devised in the AGC schemes. The linear, nonlinear models of power system can be considered as centralized, decentralized control and to get the optimized algorithms.

The research also includes the minute features, quality of load, and excitation control in AGC. AGC uses lots of methods to overcome problems, like phase shifters, HVDC (high voltage direct current) transmission links, reactive compensator, and static volt–ampere used in power control system to get the optimized results. The tiny signal is very important to get the solution for the system response. With the help of tiny signals, system can be prevented from hazardous situation. However, it is very crucial to make a counterbalance for system of nonlinearity when operating over a wide margin of operating frequency. In AGC, linearized models do not provide the desired performance when implemented with nonlinear system having the condition of large disturbances.

Due to lots of profits in technology, many researchers made it possible with the help of computational qualities and their fast performance for power systems. Fuzzy logic, GA, and neural network are the best ways and optimized methods to dodge the arduousness linked with the AGC. AGC regulators do not have sufficient knowledge to handle the problems linked with the system required to get the exact modeling. After having lots of research in AGC, there are many changes in decades, such as power industries' deregulation and uses of SMES (small and medium enterprises), photovoltaic cell (PV) cells, and wind-powered turbine, and there are other materials for electrical energy.[8]

Deregulated markets of electricity have grown up so far, having lots of multiple providers and generation and distribution, and drastic and huge changes happen at the load demand. At present, a multivariable control infrastructure is devised and a well-built design is also concocted for thermal power plant. A system with a model predictive control (MPC) approach handles the same way as a classical regulator does. When it comes to the multivariable, it corrects limitations of classical regulators. When MPC controller is not active, then SISO (Serial in Serial out) comes in contact and makes the operation go smoothly with no disconnection.[9]

Multivariable solutions are best to get the esthetic performance; the plant model is the idea to give the best possible solution for the control action. The classical model needs an operator to operate when it is necessary, manually acting on the single part of the model. Multivariable control system is not only used for optimal performance but also used for maximum malleability. Discrete-time nonlinear model is used for the production of an AGC regulator in thermal units.

The periodic changes occur when there is a drastic change in load to maintain frequency and faults in tie-lines. It is crucially important to maintain power line transfer as well as the performance of the operating frequency at both sides. This can be achieved by managing the tie-line and frequency and the power variation in system to get net difference by shrinking unstablility of different areas.[10]

There are two controlling methods: one is soft computing and another conventional control.

13.1.2 CONVENTIONAL CONTROL TECHNIQUES

13.1.2.1 INTEGRAL CONTROLLING TECHNIQUE

With the help of the frequency variation, changing of speed can be maintained naturally. After modification of the controller, the rate of the steady-state error can be minimized to zero.[10]

13.1.2.2 PROPORTIONAL WITH INTEGRAL DERIVATIVE (PID)

This is the combination of the integral and the derivative. The value changes feedback into present faults, integral modifies the feedback of new damage, and then derivative makes the whole process easy to solve by varying the faults.[10]

13.1.2.3 PROPORTIONAL INTEGRAL (PI) CONTROLLING TECHNIQUE

This controlling method is generally used universally for LFC designed systems. The major role of the PI controller is to completely remove the steady-state error to zero by giving unknown damage occurring in its previous stage.[10]

13.1.2.4 LINEAR QUADRATIC GOVERNOR–BASED SUPERVISING

Control system is basically linked in running system with low-probability cost. The condition can be shown with the help of the linear equations, and the cost can also be determined with the help of the quadratic equation.[10]

Solution of Automatic Generation Control

13.1.3 SOFT COMPUTING-BASED TECHNIQUES

13.1.3.1 ARTIFICIAL NEURAL NETWORK

By adjusting the values that are proportional to the derivative and integral terms through proportional with integral derivative (PID) controller, optimal solution is obtained.[10]

13.1.3.2 FUZZY LOGIC

FLC system mainly concocted to remove incompatibility of productivity. With the help of the reference value, all the parameters are adjusted automatically without any operator and there is no need of regular checking.[10]

13.1.3.3 GENETIC ALGORITHM

GA is the best way to get the precise outputs and this process is basically dependent on the genetics and it draws a data network that tells about the best-fit position.[10]

13.1.3.4 PARTICLE SWARM OPTIMIZATION

Particle swarm optimization method was developed by Kennedy and Eberhart in 1995, best method to get optimized results. There are lots of papers out there on the study and the performance using the nonlinear functions.[11] PSO is just like GA in which it moves in a path with some velocity according to their previous data to get the best possible solution. This search void can be any integer. Following Kennedy and Eberhart's naming convention, D is dimension of search space. ith particle can be shown as $X_i = (x_{i1}, x_{i2}, x_{i3}, ..., x_{iD})$ and previous data can be shown as $P_i = (p_{i1}, p_{i2}, ..., p_{iD})$. The whole group is g and the position change is shown as $\Delta X_i = (\Delta x_{i1}, \Delta x_{i2}, \Delta x_{i3}, ..., \Delta x_{iD})$. Every particle has the latest update regarding its position and two equations:

$$\Delta x_{id} = \Delta x_{id} + c_1 \text{rand}_1()(p_{id} - x_{id}) + c_2 \text{rand}_2()(p_{gd} - x_{id}) \quad (13.1)$$

$$X_{id} = x_{id} + \Delta x_{id} \tag{13.2}$$

Few researchers have found that c_1 and c_2 give the best performance when they are set to 2.

Originally, PSO was invented to describe only one problem at a time. However, many practical engineering problems found out that this is used for combination optimization.[11]

PSO technique is just an alternative to GA. This method is just like social animal behavior like fish in terms of schooling, bird with flapping wings, and swarm theory. In PSO, the particle contains memory so there is better chance of getting a good and optimized result but in the case of GA, there is no such memory existed in GA, the previous knowledge destroyed when population changes.[12]

The generation is the main aspect in the electrical power system. If the system collapses, then the generation leads to the hazardous issues. It can lead to the failure of electrical power system. Transmission and distribution is the most important and toughest part of the electrical power system. The technology increased the rate of the electricity. It is mandatory to use all the sources in power generation and distribution to get the full capacity of the electrical power system.[13]

The steady-state frequency error should be zero to neglect the further problems in AGC. Steady-state tie flow in the line through load change must be zero. There should be a tough controller so that there would be no wear and tear of the instruments.[14]

PSO flowchart can be explained with a simple flowchart (Fig. 13.2). At the initial position, it generates initially a set of population. It initializes the process with fitness, updates the latest information into data set, and optimizes with the best solution with the latest update. When population data set does not match the preferred values, then it creates a new set of population with having its velocity along with its position.

13.2 SIMULATION MODEL

This AGC model represents the rough idea of generation control in electrical power system. Pitch controller is used to control the angle and speed of the propeller of the turbine (Fig. 13.3).

Solution of Automatic Generation Control

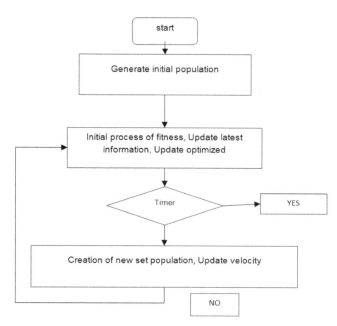

FIGURE 13.2 Flowchart of particle swarm optimization.

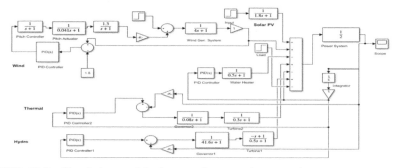

FIGURE 13.3 Simulation model of automatic generation control.

Pitch actuator: It is just a subsystem of the wind turbine, which is used to control the pitch of the control system.

Integrator: It is used to measure the signal, which is time integral of the signal.

PID controller: It is used to vary the temperature, pressure, speed, and other process variables.

Turbine-governor: It is used to store a huge amount of energy to their huge mass.

Solar PV: Solar PV is used as an input here to provide the continuous power to the system.

The future work shall be done on this model to get the best idea about the AGC.

13.3 RESULTS AND DISCUSSION

To justify the investigation part of expected algorithm, multimodal gauge functions $F1$, $F2$, $F3$, $F4$, $F5$, and $F6$ are considered, as these functions have several native optima with quantity rising aggressively w.r.t. magnitude. Table 13.1(a) presents explanation of multimodal gauge function with BMFO1 algorithm and Table 13.1(b) presents the explanation of multimodal gauge function with BMFO2 algorithm (Fig. 13.4).

TABLE 13.1(A) Results of BMFO1 Algorithm for Multimodal Gauge Function.

Benchmark functions	Parameters				
	Mean value	SD	Worst value	Best value	Wilcoxon
$F1$	−3140.2851	290.75126	−2641.0282	−4071.3905	1.73E−06
$F2$	1.6250998	0.9594445	2.9848772	0	6.87E−06
$F3$	0.038505	0.2109003	1.1551485	8.88E−16	1.01E−07
$F4$	0	0	0	0	1
$F5$	4.86E−32	1.64E−33	5.58E−32	4.71E−32	1.60E−06
$F6$	0.0010987	0.0033526	0.0109874	1.35E−32	1.20E−06

TABLE 13.1(B) Outcomes of Hybrid BMFO-SIG Algorithm for Multimodal Gauge Function.

Benchmark functions	Parameters				
	Mean value	SD	Worst value	Best value	Wilcoxon
$F1$	−3361.17	287.3247	−2879.42	−4071.39	1.73E−06
$F2$	1.392943	0.720324	2.984877	0	3.89E−06
$F3$	4.44E−15	0	4.44E−15	4.44E−15	4.32E−08
$F4$	0	0	0	0	1
$F5$	4.82E−32	8.59E−34	5.12E−32	4.71E−32	1.56E−06
$F6$	0.002558	0.010248	0.054779	1.35E−32	1.34E−06

Solution of Automatic Generation Control

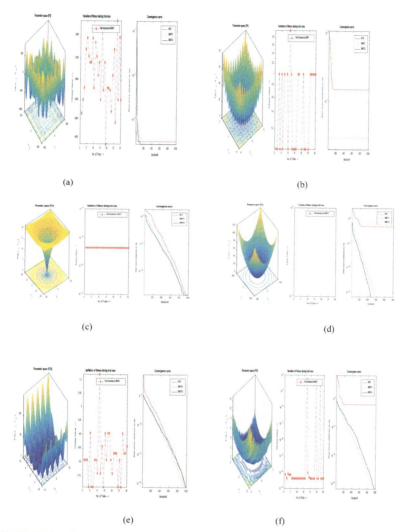

FIGURE 13.4 Convergence curve and trial solutions of BMFO2 for multimodal gauge functions.

13.4 CONCLUSION

In the proposed analysis, two binary variants of moth-flame optimizer have been given to resolve benchmark issues and AGC downside of the electrical power grid. Results of two variants have been proved for nonprecise, extremely unnatural, nonarched engineering style and optimization issues that embody 23 benchmark issues and AGC downside of multispace power grid below deregulating setting. The planned algorithm has been accustomed to examine the optimum gain of PI controller with suitable error criteria. Moreover, the efficiency of BMFO algorithm tuned with PI controller is additionally correlated with standard MFO and PSO formula tuned with PI controller in several contract eventualities of a deregulated power grid. The analyses disclosed that proposed BMFO algorithm offers superior kind of solutions as correlated to alternatively described metaheuristics search algorithms. In future work, the effectualness of the BMFO formula is studied for the optimum answer to the many alternative engineering issues.

KEYWORDS

- **automatic generation control**
- **nonconventional energy sources**
- **particle swarm optimization**
- **load frequency control**
- **multi area power plant**

REFERENCES

1. Abd Elaziz, M.; Oliva, D.; Xiong, S. An Improved Opposition-Based Sine Cosine Algorithm for Global Optimization. *Expert Syst. Appl.* **2017**, *90*, 484–500.
2. Wolpert, D. H.; Macready, W. G. No Free Lunch Theorems for Optimization. *IEEE Trans. Evol. Comput.* **1997**, *1* (1), 67–82.
3. Simon, D. Biogeography-Based Optimization. *IEEE Trans. Evol. Comput.* **2008**, *12* (6), 702–713.
4. Mirjalili, S.; Mirjalili, S. M.; Lewis, A. Grey Wolf Optimizer. *Adv. Eng. Softw.* **2014**, *69*, 46.
5. Mirjalili, S. The Ant Lion Optimizer. *Adv. Eng. Softw.* **2015**, *83*, 80–98.
6. Mirjalili, S. Moth-Flame Optimization Algorithm: A Novel Nature-Inspired Heuristic Paradigm. *Knowledge Based Syst.* **2015**, *89*, 228–249.
7. Mirjalili, S.; Mirjalili, S. M.; Hatamlou, A. Multi-Verse Optimizer: A Nature-Inspired Algorithm for Global Optimization. *Neural Comput. Appl.* **2016**, *27* (2), 495–513.

8. Mirjalili, S. Dragonfly Algorithm: A New Meta-Heuristic Optimization Technique for Solving Single-Objective, Discrete, and Multi-Objective Problems. *Neural Comput. Appl.* **2016,** *27* (4), 1053–1073.
9. Mirjalili, S. SCA: A Sine Cosine Algorithm for Solving Optimization Problems. *Knowledge Based Syst.* **2016,** *96,* 120–133.
10. Shareef, H.; Ibrahim, A. A.; Mutlag, A. H. Lightning Search Algorithm. *Appl. Soft Comput. J.* **2015,** *36,* 315–333.
11. Chaohua, D.; Weirong, C.; Yunfang, Z. Seeker optimization algorithm. *2006 Int. Conf. Comput. Intell. Secur. ICCIAS 2006,* **2007,** *1,* 225–229.
12. Li, M. D.; Zhao, H.; Weng, X. W.; Han, T. A Novel Nature-Inspired Algorithm for Optimization: Virus Colony Search. *Adv. Eng. Softw.* **2016,** *92,* 65–88.
13. Mirjalili, S.; Lewis, A. The Whale Optimization Algorithm. *Adv. Eng. Softw.* **2016,** *95,* 51–67.
14. Bayraktar, Z.; Komurcu, M.; Werner, D. H. Wind Driven Optimization (WDO): A Novel Nature-Inspired Optimization Algorithm and Its Application to Electromagnetics. *2010 IEEE Int. Symp. Antennas Propag. CNC-USNC/URSI Radio Sci. Meet.—Lead. Wave, AP-S/URSI 2010* **2010,** *1,* 0–3.

CHAPTER 14

IOT-Based Technology for Smart Farming

VIKALP JOSHI[1] and MANOJ SINGH ADHIKARI[2*]

[1]Department of ASS, Sara Sae PVT Limited, Dehradun, Uttarakhand, India

[2]School of Electronics & Electrical Engineering, Lovely Professional University, Phagwara, Punjab, India

*Corresponding author. E-mail: manoj.space99@gmail.com

ABSTRACT

In modern time, Internet of thing (IoT) plays major roles to provide precise, accurate data and information for agriculture industry and farmers. Such innovative modifications are mixing the established agricultural techniques and making novel chance along a range of challenges. Using the arrangement of progressive technologies in hardware and software, the IoT is capable to track and sum everything which can importantly reduce the waste. The sensors like moisture, soil preparation, irrigation, and rainfall are the backbone of IoT. This chapter deals with the capability of IoT and nanotechnologies-based wireless sensors generally known as mote.

14.1 INTRODUCTION

To gain the agricultural revenue with very less resource and labor efforts, the Internet of thing (IoT) can be used in agriculture industries.[1–5] IoT-based technologies provide facilities to farmers and agriculture industries to check the status of crops time to time without visiting the worksite.[6–10] The accurate, precise solution and advance technologies are extremely needed in agriculture industry due to its vastness.[11–14] The farmers can observe the status of the worksite by using nanotechnologies-based smart sensors and wireless communication system. The modern agricultural technologies are based on smart apparatus and kits used for insemination to crop mowing and even during storage space and shipping. By periodical update through historical trend and graphs, IoT-based agricultural technique becomes smart and cost effective. Embedded technologies-based drones, automotive tractors, harvesters, and satellites are the backbone of the IoT-based agricultural system.[15–22] A few years ago, IoT-based technologies provide facilities in manufactured, medical, communication, energy, and many more sectors to reduce inefficiencies and improve product qualities.[23–30] Figure 14.1 shows the main key points of technology while Figure 14.2 shows obstacle of technology execution in smart agriculture.

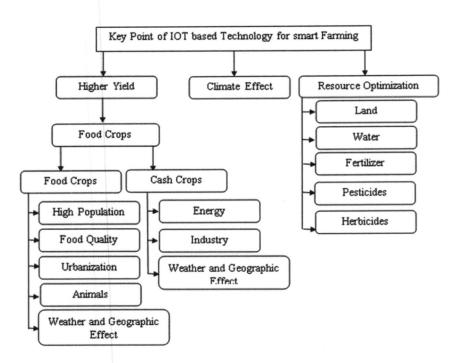

FIGURE 14.1 Main key points of technology.

IOT-Based Technology for Smart Farming 225

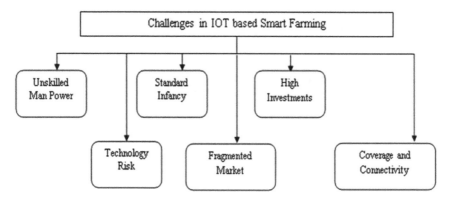

FIGURE 14.2 Obstacle of technology execution in smart agriculture.

To monitor and control the use of IoT-based agricultural technologies to regulate the sustaining food and environmental safety, Food Corporation of India (FCI), several agricultural commissions, and government authorities provide time-to-time polices and guidelines. This chapter represents and evaluates the trends of IoT-based agricultural technologies in order to develop crop quality and production.[31–38]

There are important points of the IoT for the industry-related agriculture in the chapter:

- potential of the land of the earth from the agricultural sector,
- latest research and development on agricultural industries in universities and industry levels,
- restrictions of agriculture industries,
- role of IoT in agriculture industry like monitor and control resource shortage, food damage. Weather and environmental pollution status, and
- important unsolved issues and suggestions.

This chapter is useful for engineers and researchers to provide smart agricultural technologies through the power of IoT. The next section provides a major application of IoT in agricultural sector and what can be possible by IoT.[39–44]

14.2 MAJOR APPLICATIONS

Conventional farming technique can be totally changed by implementation of high precise quality sensor and communication-based IoT techniques. Presently, consolidated wireless integration sensors technique increased the level of agricultural technique. By continuous improvement of agricultural technology, several farming issues like yield optimization, drought response, land suitability, pest control, and irrigation can be improved. Figure 14.3 shows a hierarchy of high precise quality sensors, service, and applications for smart agriculture.

14.2.1 MAPPING AND SAMPLING OF SOIL

Soil is the heart of the plant. To determine the worksite information that is useful to decide the crop condition at different levels, sampling is necessary. Soil analysis is used to find the amount of nutrient. This analysis is very useful to find nutrient deficiencies. Based on worksite conditions, a comprehensive soil test is performed on an annual basis. The soil quality, cropping history, fertilizer application, and level of irrigation are critical factors to analyze the soil. The previous analysis is very useful to check the biological, physical, and chemical status of soil. Generally, soil mapping is useful for selection of seed suitability, sowing time, and depth of sowing for different types of crops.

Presently, several sensors and toolkits are available in market to identify the status and quality of soil nutrient. This is very helpful for farmers to avoid degradation. IoT-based agricultural technique can monitor moisture contain absorption, texture, and rate of absorption that are very useful to minimize acidification, pollution, and intersection. Crop productivity depends on several factors, but one important factor is soil dryness. The farmers in the world face soil dryness problem. To overcome soil dryness problem, measurement of soil moisture through remote sensing technique is used. This technique is useful for soil dryness in remote regions. The Satellite for Soil Moisture and Ocean Salinity was launched early in 2009, which provide maps of moisture of soil of remote regions day by day.[1]

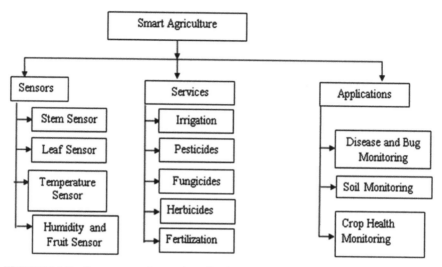

FIGURE 14.3 Sensors, service, and application of IoT-based smart farming.

In these approaches, to determine the soil moisture, field data of soil moisture is compared to water deficit index (SWDI).[2] In this offer, they followed a dissimilar technique to determine the soil moisture in order to compare with the SWDI acquired from in situ data. The prediction model is based on field survey data and soil mapping.

To fix the depth and location for insemination the seed, real-time camera and sensors-based techniques are used.[3] The robots record the field data and transmit to master control station where the data can be analyzed. The different contactless sensing techniques are planned to decide the flow rate of seed. Such type of sensors are attached visible and laser LED, transmitter receiver, and battery backup.[4] The movement of the seed is captured through the sensors and light detector where its signal conditioning system converts sensor signal into transmission signal form.

14.2.2 IRRIGATION

In total, only 3% of the total water on the earth is in fresh form and the remaining 97% is salty sea and ocean water. Two-thirds of water is in the form of glaciers and frozen form and 0.5% of water found in the form of groundwater.[5] Summing up, human life totally depends on 0.5% of water to accomplish all its necessities and to sustain the ecosystem; therefore, the sufficient fresh water is reserved in lakes, rivers, and water resources to maintain it. Generally, the 70% of accessible fresh water is required to drive agricultural sector.[7]

In several countries like Brazil the water consumption increases to 75% and in a developing countries it exceeds to 80%.[8] The major cause for this large water expenditure is observed in 2013, and physical inspection of crops is necessary for irrigation purpose. Approximately 80% of farms in United States used such type of technique.[9] According to UN Convention to Combat Desertification, the half of the world can be fully affected by water shortage problem by 2030. And around 168 countries are affected by desertification.[10] Sprinkler and drip irrigation–type different techniques are used to initiate waste water issues that were also established in conventional techniques.

When water shortage problem occurs, the quantity and quality of crops are severely affected. If irrigation is not regular, decrease in soil nutrients and increase in diverse microbial infections occur. This is not an easy task to precisely guess the water requirement of crops where factors similar to precipitation, soil moisture, crop category, irrigation technique, soil variety, and crop requirements are involved. The present condition of irrigation technique is likely to be altered by implementing the emergent IoT technologies.

IoT-based techniques play a very significant role to improve crop efficiency like irrigation management system based on crop water stress index (CWSI).[9] To calculate CWSI, crop canopy at dissimilar time interval and air temperature is recorded.

For the previously mentioned measurement parameters, the intelligent software applications are used to connect through wireless sensors monitoring system. For water need assessment in CWSI model, climate condition statistics and satellite imaging are required. To improve efficiency of water utilization, variable rate irrigation optimization technique is used.[11]

14.2.3 FERTILIZER

To improve growth and fertility of plants, fertilizers (consisting natural or chemical substance) are used. The three macronutrients of plants potassium, nitrogen, and phosphorus are used for root, flowers, leaf, and fruit development.[12] Lack of nutrients can be harmful for the plant growth. Approximately one-fourth nitrogen in the form of fertilizer is absorbed by crop. The remaining portion of fertilizer is emitted to the atmosphere. The soil nutrient level is unbalanced due to uneven use of fertilizer.

14.2.4 MANAGEMENT OF PEST AND CROP DISEASE

The 20–40% of crop yields are lost due to pests and diseases as estimated by Food and Agriculture Organization. Agriculture-related chemicals and pesticides became an essential element of the agriculture industry. The pesticides are not good for human and animals. Wireless sensors, robots and drones, and latest IoT-based intelligent systems are used to control damage. To control pest in conventional time, calendar or recipe are used, but in modern time, IoT-based technique provides real-time monitoring and forecasting.[15]

14.2.5 MONITORING FORECASTING AND HARVESTING OF YIELD

To examine a range of aspects related to agricultural yield is known as yield monitoring, like mass flow of grain, amount of moisture, and quantity of harvested grain. This technique is very useful to access by recording the moisture and crop level to guess how much quality the crop achieved and what to do ahead. Yield quality depends on several factors.[16,17]

14.3 ADVANCED AGRICULTURAL PRACTICES

To maintain the high rate of urban population, Section 14.3 shows the function of IoT in modern agricultural sector similar to vertical farming, phenol typing, and hydroponics. The implementation of IoT in agricultural sector by using advanced technologies like sensor robots, tractors, and wireless device is described in Section 14.4. Section 14.5 shows the advantages of IoT technologies. Sections 14.6 and 14.7 indicate the application of IoT to guarantee food quality for long time and major challenges to IoT technologies and to end starving. Section 14.8 represents the conclusion of IoT-based technologies in agricultural sector.

To implement the new technique in agricultural sector and to improve the superiority and production of crop are not somewhat innovative, as farmers have been performing this task for eras. Primarily, we are working to improve the seed quality, fertilizers, and pesticides to improve crop production.

14.3.1 GREENHOUSE FARMING

This type of farming is treated as the firstborn technique of smart farming. The oldest method of smart farming is greenhouse technique, although the suggestion of raising plants in organized surroundings is not novel and established since ancient times. The highest production rate of numerous crops under such well-ordered atmosphere depends on several features, like correctness of nursing parameters, arrangement of ventilation system, material to control wind effects, decision support system, and shed.

14.3.2 VERTICAL FARMING

The too much farmable land required fulfilling increased food demands; however, due to erosion and pollution, one-fourth arable property was vanished through the previous four decades.[18,19] Unfortunately, the industrial farming used in modern agricultural technique is very harmful to soil quality and can only be rebuilt through nature.

14.3.3 HYDROPONIC

This is the subset of hydro culture. To improve the profit of greenhouse farming, the agricultural professionals moved ahead additional steps, providing the plan of hydroponic. Hydro culture is a method through which floras are developed without soil. In this technique, roots of the crop kept in the water and some balance nutrients are added to them. This technique is also known as hydroponic.

14.3.4 PHENOTYPING

This technique is more reliable to produce good quality and quantity of crop. Some advanced methods are underneath examination to additional crop abilities and their limitation with the deployment of wireless and sensing technologies. The wireless communication and remote sensing technique are used in modern harvesters, tractors, and different types of robot in large-scale agriculture. The GPS and GIS facilities–based tractors, harvesters, and different types of robots are helpful in precision agriculture. The success rate of exactness agriculture is depending on accurateness of sample data. Generally success rate depends in two ways.[20]

14.3.5 WIRELESS SENSORS

Presently, different types of the apparatus for smart farming are available in the market. To monitor the crop status, wireless sensors play the most important role. Wireless sensors are being used by individuals wherever necessary and associated to tools and heavy equipment to perform different tasks for a particular application. The purpose and working procedure of some sensors are discussed later.

14.3.5.1 ACOUSTIC SENSORS

This category of sensors are used in different types of appliances like farm management. Generally, farm management includes cultivation of soil, weeding, and fruit harvesting, and the acoustic sensors provide fast response with low cost. The basic concept of acoustic sensors is for the measurement of level of noise when the input sensing element is attached to other material, for example, soil particle.[21] To monitor pests and detect seed qualities, acoustic sensors are used.

14.3.5.2 FIELD-PROGRAMMABLE GATE ARRAY–TYPE SENSORS

These types of sensors provide the reconfiguration flexibility. FPGA (Field Programmable Gate Array) sensors measure plant transpiration, humidity, and irrigation in real time. The major disadvantages of FPGA are not suitable in real-time monitoring and require more power.

14.3.5.3   OPTICAL SENSORS

This type of sensors is very helpful in agriculture purpose due to low cost, ease of use, and adjustability. Optical sensors are used to measure capability of soil with the help of electromagnetic spectrum concept. Optical sensor reflects light, and the changes appearing in wave reflections assist to show the variations in density of soil and several additional parameters. The optical sensor is associated to microwave scattering purpose and can be used to detect quality of grove canopies and similar related crops.

14.3.5.4   ULTRASONIC-TYPE RANGING SENSORS

This type of sensors are of low cost, used in different locations and conditions, and have easy installation process. The sensors are useful for tank monitoring, spraying, and distance measurement. Camera-based ultrasonic ranging sensors are used for weed detection and crop coverage purpose.[22] Actually, ultrasonic sensors measure plant length, and the camera shows the forest grass and crop coverage.

14.3.5.5   OPTOELECTRONIC-TYPE SENSORS

This sensor is helpful to detect the type of plant, weeds, and herbicides.[23] Sensor data and location information are useful for mapping weed distribution and resolution.[24]

14.3.5.6   AIRFLOW SENSORS

For the measurement of soil air permeability and moisture, the detection of soil structure, and for differentiating soil categories, airflow-type sensors are used.

14.3.5.7 ELECTROCHEMICAL SENSORS

Electrochemical sensors are typically used to evaluate the important soil characteristics to examine the amount of soil nutrient like pH.[25] These types of sensors replace expensive and time-consuming soil chemical analysis process (Table 14.1).

TABLE 14.1 Sensors and their Application in IoT-Based Agriculture Sector.

S. no	Sensor	System sensor to be attached	Measurement variable
1.	XH-M214[26]	Equipment	Yield and moisture
2.	PYCNO[27]	Soil and weather	Water
3.	Sol Chip Com[29]	Soil	Pollution and water
4.	MP 406[28]	Soil	Temperature and moisture
5.	DEX70[30]	Plant	Fruit size
6.	Wind Sentry 03002[31]	Weather	Wind
7.	SF-4/5[32]	Plant	Fruit size
8.	Met Station One[33]	Weather	Wind
9.	SD-6P[34]	Plant	Fruit size
10.	B-102[35]	Equipment	Location tracking

14.3.6 IOT-BASED TRACTORS

The development of the crop industry creates problem for rural labors because industry apparatus and equipment like tractors and heavy machine work 40 times faster as compared to a human being. To accomplish the continuously growing demand, agricultural apparatus and equipment like John Deere and Hello Tractor have initiated to provide superior solutions.

14.3.7 HARVESTING ROBOTS

The final stages of production process are harvesting. Crop harvesting is one of the important parameters at which the time is very significant, so it can affect the production considerably.

14.3.8 COMMUNICATION IN AGRICULTURE

The backbone of precision agriculture is to provide the communication and reporting the field data periodically. To provide field data periodically, telecom providers can play a major role in the agriculture.

14.3.8.1 CELLULAR COMMUNICATION

To collect the field data, the second-generation to fourth-generation communication can be suitable. The communication depends on bandwidth requirement (Table 14.2).

TABLE 14.2 Application and Specification of Wireless Sensors.

S. no	Types of data	Application	Data size (approx.)	Power consumption
1.	Small size	Air, soil temperature, humidity, leaf thickness	100 B/s	10 mA
2.	Medium size	Picture camera, multi-spectral camera, acoustic camera	10 Mb/s	100 mA
3.	Large size	Video-streaming cameras	10 Mb/min	50 A

14.3.8.2 ZIGBEE

Zigbee is mainly used in a large range of purposes, particularly to substitute prevailing nonstandard technologies. Generally, Zigbee is used to monitor greenhouse atmosphere and short-ranged communication purpose.

14.3.8.3 BLUETOOTH

This is useful when worksite and data collection center are not very far. To commence data communication between small devices, wireless communication standard is used. The Bluetooth-based moisture and temperature sensors monitor the impact of environment and weather on the crops in worksite. A similar type of sensors are developed which are known as nodes that are useful for the measurement of light intensity and surrounding temperature.

14.3.8.4 LORA

This type of technologies are used for a long range of purposes and also where low power is required in IoT-based applications.

14.3.8.5 SIGFOX

SIGFOX is low-powered communication system.

Generally, to inform most of the farming groups of people, the primary and common source of communication in rural areas is mobile phones. Modern advancements in the cellular phone sector have resulted in intense significance, making agricultural sector much attractive, particularly to the smallholders in remote areas (Table 14.3).

TABLE 14.3 Sensor Based on Smart Phone Used in Agriculture Applications.

S. no.	Smart phone sensor	Motto	Application in agricultural sector
1.	Real-time camera	Capture pictures of any object	Plant disease analysis, soil erosion, leaf area analysis purpose
2.	GPS	Location detection, monitoring the latitude and longitude of device	Site information is attached to create alerts. Generally used for machine driving and tracking, land management purpose, and crop mapping
3.	Microphone	Detects wanted/unwanted sound and convert to electrical signals	Maintenance of machine, detection of bug
4.	Accelerometer	Measurement of acceleration forces	Monitor activities of worker or machine activities
5.	Gyroscope	Measurement of angular velocity	Movement of equipment
6.	Barometer	Measurement of air pressure as an altimeter	Measurement of the elevation height in hilly agriculture

14.3.9 CLOUD COMPUTING

Cloud computing is very useful in precision agriculture. Figure 14.4 represents the fluid computing for smart farming.

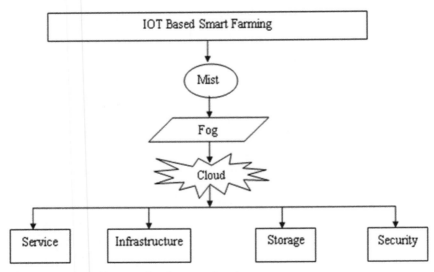

FIGURE 14.4 Fluid computing for smart farming.

14.4 UNMANNED AERIAL VEHICLES

In modern times, IoT creates remarkable improvement in several industries similar to poultry and fishing in farming sector. This type of communication infrastructure and facilities are rarely available in undeveloped countries and remote areas, which is one of the difficulties to introduce IoT in the agricultural sector.

The drones play a major role for monitoring the crop. Some of the key points of drone which are used in agricultural sector are given later.

14.4.1 SOIL AND WORK SITE ANALYSIS

To examine the soil conditions before the harvest, drones cameras are helpful to produce precise and accurate information that assists to decide the most appropriate time of harvest for a particular plot. Furthermore, it also suggests the category of seed and design of planting.

14.4.2 PLANTING

Presently, lots of acres of property are not used due to human inaccessibility.

14.4.3 MONITORING OF CROP

To monitor the crop in large covering area is a tough job. Drones cameras provide more accurate and cost-effective images in real time.

14.4.4 IRRIGATION

IoT-based drones are also used for irrigation purpose. It consists of twofolds; one side associated unmanned aerial vehicles (UAV) with different types of sensors and camera. This type of technique captures the image, field data of crop, and at the same time is used for water sprinkle purpose. The technique reduces working time and is very useful in emergency time period.

14.4.5 DETECTION OF GAP AND PLANT COUNTING

Precision agriculture importantly inclines the 3D data on crop density when creating choices throughout different types of applications.

14.4.6 SPRAYING THE PESTICIDES/HERBICIDES

To spray pesticides on crops, UAV can be used. However, implementing such type of applications is more critical.

14.4.7 HEALTH ASSESSMENT

The IR (infrared) light sensors and drones can recognize crops that are infected by bacteria or fungus.

14.4.8 RECOGNITION OF PLANT SPECIES

To monitor the plant species specifically which are vanished from earth, UAV play major role to detect and recognize them. A UAV have the ability to monitor plant species from inaccessible areas. Another scope of UAV is to detect the forest biomass and fuel with the help of radars and satellites.

14.5 PRESENT CHALLENGES AND FUTURE EXPECTATIONS

The UN and international community decide in 2015 "The 2030 Agenda for Sustainable Development," to end starvation by 2030, although by the recent facts shown by WHO, approximately 800 million people are facing the food shortage problem. The statistics are alarming on their own. Another serious issue is food quality that is more critical. The resources are available in our planet, but it is in our hands to learn how to use them wisely and accurately. If we use technologies in sensible manner then the possibility is definitely there to remove starvation in upcoming decades. To remove starvation, there is a need to start such organizations and institutes to provide policies for utilization of technologies in sensible manner.

14.6 CONCLUSION

IoT-based agricultural technologies can improve crop growing, quality, and production rate in order to maintain upcoming food demand. These techniques have sufficient capabilities to end starvation challenges. A research conducted in 195 countries on "effect of unhealthy diet" from 1990 to 2017 showed that death rate in one out of five per year could be clog by given that healthier diet. The responsible risk factor for deaths is taking low grains diet. In many regions, farmers are following the conventional agricultural technique while trying to conjoin the food requirements by better utilizing fertilizers and pesticides. The blind uses of fertilizers and pesticides create irreversible implications to the environment. The agricultural research organization and farmers need to run differently and use modern technologies to improve crop quality and quantity.

KEYWORDS

- IoT
- sensors
- FCI
- FPGA

REFERENCES

1. Martínez, F. J.; González, Z. A.; Sánchez, N.; Gumuzzio, A.; Herrero, J. C. M. Satellite Soil Moisture for Agricultural Drought Monitoring: Assessment of the SMOS Derived Soil Water Deficit Index. *Remote Sens. Environ.* **2016**, *177*, 210–220.
2. Vågen, T. G.; Winowiecki, L. A.; Tondoh, J. E.; Desta, L. T.; Gumbricht, T. Mapping of Soil Properties and Land Degradation Risk in Africa Using MODIS Reflectance. *Geoderma* **2016**, *263*, 216–225.
3. Santhi, P. V.; Kapileswar, N.; Chenchela, V. K. R.; Prasad, C. H. V. S. Sensor and Vision Based Autonomous AGRIBOT for Sowing Seeds. *Int. Conf. Energy Commun. Data Anal. Soft Comput.* **2017**, *1*, 100–105.
4. Karimi, H.; Navid, H.; Besharati, B.; Behfar, H.; Eskandari, I. A Practical Approach to Comparative Design of Non-contact Sensing Techniques for Seed Flow Rate Detection. *Comput. Electron. Agric.* **2017**, *142*, 165–172.
5. Romero-Trigueros, C.; Nortes, P. A.; Alarcón, J. J.; Hunink, J. E.; Parra, M.; Contreras, S.; Droogers, P. Effects of Saline Reclaimed Waters and Deficit Irrigation on Citrus Physiology Assessed by UAV Remote Sensing. *Agric. Water Manage.* **2017**, *183*, 100–105.
6. Hoffmann, H.; Jensen, R.; Thomsen, A.; Nieto, H.; Rasmussen, J.; Friborg, T. Crop Water Stress Maps for an Entire Growing Season from Visible and Thermal UAV Imagery. *Biogeosciences* **2016**, *13*, 6545–6563.
7. Andriamandroso, A. L. H.; Lebeau, F.; Beckers, Y.; Froidmont, E.; Dufrasne, I.; Heinesch, B.; Dumortier, P.; Blanchy, G.; Blaise, Y.; Bindelle, J. Development of an Open-Source Algorithm Based on Inertial Measurement Units (IMU) of a Smartphone to Detect Cattle Grass Intake and Ruminating Behaviors. *Comput. Electron. Agric.* **2017**, *139*, 126–137.
8. Azam, M. F. M.; Rosman, S. H.; Mustaffa, M.; Mullisi, S. M. S.; Wahy, H. Jusoh, M. H.; Ali, M. I. M. Hybrid Water Pumps System for Hilly Agricultural Site. *IEEE Control Syst. Grad. Res. Colloquium (ICSGRC).* **2017**, *1*, 100–105.
9. Hunter, M. C.; Smith, R. G.; Schipanski, M. E.; Atwood, L. W.; Mortensen, D. A. Agriculture in 2050: Recalibrating Targets for Sustainable Intensification. *Bio Sci.* **2017**, *67* (4), 386–391.
10. Qian, J.; Yang, X.; Wu, X.; Xing, B.; Wu, B.; Li, M. Farm and Environment Information Bidirectional Acquisition System with Individual Tree Identification

Using Smartphones for Orchard Precision Management. *Comput. Electron. Agric.* **2015**, *1*, 101–108.
11. Debauche, O.; Mahmoudi, S.; Andriamandroso, A. L. H.; Manneback, P.; Bindelle, J.; Lebeau, F. Cloud Services Integration for Farm Animals Behavior Studies Based on Smartphones as Activity Sensors. *J. Ambient Intell. Humaniz. Comput.* **2018**, *1*, 1–12.
12. Orlando, F.; Movedi, E.; Coduto, D.; Parisi, S.; Brancadoro, L.; Pagani, V.; Guarneri, T.; Confalonieri, R. Estimating Leaf Area Index (LAI) in Vineyards Using the Pocket LAI Smart-App. *Sensors (Basel)* **2016**, *16*, 1–12.
13. Frommberger, L.; Schmid, F.; Cai, C. Micro-mapping with Smart Phones for Monitoring Agricultural Development. *ACM Dev.* **2013**, *1*, 100–105.
14. Jin, X.; Liu, S.; Baret, F.; Hemerlé, M.; Comar, A. Estimates of Plant Density of Wheat Crops at Emergence from Very Low Altitude UAV Imagery. *Remote Sens. Environ.* **2017**, *198*, 105–114.
15. Venkatesan, R.; Kathrine, G.; Jaspher, W.; Ramalakshmi, K. Internet of Things Based Pest Management Using Natural Pesticides for Small Scale Organic Gardens. *J. Comput. Theor. Nanosci.* **2018**, *15*, 100–104.
16. Wietzke, A.; Westphal, C.; Gras, P.; Kraft, M.; Pfohl, K.; Karlovsky, P.; Pawelzik, E.; Tscharntke, T.; Smit, I. Insect Pollination as a Key Factor for Strawberry Physiology and Marketable Fruit Quality. *Agric. Ecosyst. Environ.* **2018**, *258*, 100–105.
17. Dehaghi, M. A. Effects of Biological and Chemical Fertilizers Nitrogen on Yield Quality and Quantity in Cumin. *J. Chem. Health Risks* **2014**, *4* (2), 55–64.
18. Kou, Z.; Wu, C. Smartphone Based Operating Behavior Modeling of Agricultural Machinery. *IFAC-PapersOnLine* **2018**, *51* (17), 521–525.
19. Machado, B. B.; Orue, J. P. M.; Arruda, M. S.; Santos, C. V.; Sarath, D. S.;. Goncalves, W. N.; Silva, G. G.; Pistori, H.; Roel, A. R.; Rodrigues-Jr, J. F. BioLeaf: A Professional Mobile Application to Measure Foliar Damage Caused by Insect Herbivore. *Comput. Electron. Agric.* **2016**, *129*, 44–55.
20. Zhang, L.; Dabipi, I. K.; Brown, W. L. Internet of Things Applications for Agriculture. In *Internet of Things A to Z: Technologies and Applications*; Hassan, Q., Ed.; 2018.
21. Kong, Q.; Chen, H.; Mo, Y. L.; Song, G. Real-Time Monitoring of Water Content in Sandy Soil Using Shear Mode Piezo-ceramic Transducers and Active Sensing—A Feasibility Study. *Sensors* **2017**, *2395*, 100–105.
22. Pajares, G.; Peruzzi, A.; Gonzalez-de-Santos, P. Sensors in Agriculture and Forestry. *Sensors (Basel)* **2013**, *1*, 12132–12139.
23. Andújar, D.; Ribeiro, Á.; Fernández-Quintanilla, C.; Dorado, J. Accuracy and Feasibility of Optoelectronic Sensors for Weed Mapping in Wide Row Crops. *Sensors* **2011**, *1*, 2304–2318.
24. Xie, X.; Zhang, X.; He, B.; Liang, D.; Zhang, D.; Huang, L. A System for Diagnosis of Wheat Leaf Diseases Based on Android Smartphone. *Opt. Meas. Technol. Instr.* **2016**, *10155*, 100–105.
25. Yew, T. K.; Yusoff, Y.; Sieng, L. K.; Lah, H. C.; Majid H.; Shelida, N. In *An Electrochemical Sensor ASIC for Agriculture Applications*, International Convention on Information and Communication Technology, Electronics and Microelectronics (MIPRO); Opatija, 2014; pp 85–90.

26. Maldonado, W.; Valeriano, T. T. B.; Souza Rolim, G. EVAPO: A Smartphone Application to Estimate Potential Evapotranspiration Using Cloud Gridded Meteorological Data from NASA-POWER System. *Comput. Electron. Agric.* **2019**, *156*, 187–192.
27. Bartlett, A. C.; Andales, A. A.; Arabi, M.; Bauder T. A. A Smartphone App to Extend Use of a Cloud-Based Irrigation Scheduling Tool. *Comput. Electron. Agric.* **2015**, *111*, 127–130.
28. Freebairn, D.; Robinson, B.; Mcclymont, D.; Raine, S.; Schmidt, E.; Skowronski, V.; Eberhard, J. In *Soil Water App-Monitoring Soil Water Made Easy*, Proc.18th Aust. Soc. Agron Conf., Sept 24–28, 2017.
29. Ferguson, J. C.; Chechetto, R. G.; O'Donnell, C. C.; Fritz, B. K.; Hoffmann, W. C.; Coleman, C. E.; Chauhan, B. S.; Adkins, S. W.; Kruger, G. R.; Hewitt, A. J. Assessing a Novel Smartphone Application-SnapCard, Compared to Five Imaging Systems to Quantify Droplet Deposition on Artificial Collectors. *Comput. Electron. Agric.* **2016**, *128*, 193–198.
30. Jordan, R.; Eudoxie, G.; Maharaj, K.; Belfon, R.; Bernard, M. Agri Maps: Improving Site-specific Land Management Through Mobile Maps. *Comput. Electron. Agric.* **2016**, *123*, 100–105.
31. Bueno-Delgado, M. V.; Molina-Martínez, J. M.; Correoso Campillo, R.; Pavón-Mariño, P. Ecofert: An Android Application for the Optimization of Fertilizer Cost in Fertigation. *Comput. Electron. Agric.* **2016**, *121*, 32–42.
32. Sopegno, A.; Calvo, A.; Berruto, R.; Busato, P.; Bocthis, D. A Web Mobile Application for Agricultural Machinery Cost Analysis. *Comput. Electron. Agric.* **2016**, *130*, 158–168.
33. Palomino, W.; Morales, G.; Huaman, S.; Telles, J. In *PETEFA: Geographic Information System for Precision Agriculture*, IEEE Conf. Electron. Electr. Eng. Comput.; INTERCON, 2018; pp 1–4.
34. Herrick, J. E. The Land-Potential Knowledge System (Landpks): Mobile Apps and Collaboration for Optimizing Climate Change Investments. *Ecosyst. Health Sustain.* **2016**, *2*, 100–105.
35. Nakato, G. V.; Beed, F.; Bouwmeester, H.; Ramathani, I.; Mpiira, S.; Kubiriba, J.; Nanavati, S. Building Agricultural Networks of Farmers and Scientists via Mobile Phones: Case Study of Banana Disease Surveillance in Uganda. *Can. J. Plant Pathol.* **2016**, *38*, 307–316.
36. Masuka, B.; Matenda, T.; Chipomho, J.; Mapope, N.; Mupeti, S.; Tatsvarei, S. Mobile Phone Use by Small-Scale Farmers: A Potential to Transform Production and Marketing in Zimbabwe. *S. Afr. J. Agric. Extension* **2016**, *44* (2), 121–135.
37. Chung, S.; Breshears, L. E.; Yoon, J. Y. Smartphone Near Infrared Monitoring of Plant Stress. *Comput. Electron. Agric.* **2018**, *154*, 93–98.
38. McGonigle, A. J. S.; Wilkes, T. C.; Pering, T. D.; Willmott, J. R.; Cook, J. M.; Mims, F. M.; Parisi, A. V. Smartphone Spectrometers. *Sensors (Switzerland)* **2018**, *18* (1), 100–105.
39. Moonrungsee, N.; Pencharee, S.; Jakmunee, J. Colorimetric Analyzer Based on Mobile Phone Camera for Determination of Available Phosphorus in Soil. *Talanta* **2015**, *136*, 204–209.

40. Camacho, H. A. A. Smartphone-Based Application for Agricultural Remote Technical Assistance and Estimation of Visible Vegetation Index to Farmer in Colombia: Agro TIC. *Remote Sens. Agric. Ecosyst. Hydrol. SPIE Remote Sens.* **2018,** *10783,* 100–105.
41. Prosdocimi, M.; Burguet, M.; Di Prima, S.; Sofia, G.; Terol, E.; Rodrigo Comino, J.; Cerdà, A.; Tarolli, P. Rainfall Simulation and Structure-from-Motion Photogrammetry for the Analysis of Soil Water Erosion in Mediterranean Vineyards. *Sci. Total Environ.* **2017,** *574,* 204–215.
42. Han, P.; Dong, D.; Zhao, X.; Jiao, L.; Lang, Y. A Smartphone-Based Soil Color Sensor: For Soil Type Classification. *Comput. Electron. Agric.* **2016,** *123,* 232–241.
43. Yang, Y.; Wan, X.; Cui, J.; Zheng, T.; Jiang, X.; Zhang, J. Smartphone Based Hemispherical Photography for Canopy Structure Measurement. *Optoelectron. Meas. Technol. Syst.* **2018,** 1–7.
44. Minkoua Nzie, J. R.; Bidogeza J. C.; Ngum N. A. Mobile Phone Use Transaction Costs, and Price: Evidence from Rural Vegetable Farmers in Cameroon. *J. Afr. Bus* **2017,** 1–20.

CHAPTER 15

Orthogonal Frequency Division Multiplexing for IoT

ARVIND KUMAR* and RAJOO PANDEY

[1]Department of Electronics & Communication Engineering, NIT, Kurukshetra, India

*Corresponding author. E-mail: arvind_sharma@nitkkr.ac.in

ABSTRACT

OFDM systems have high spectral efficiency because of overlapping spectra of subcarriers and may be used in 5G NR techniques, IoT applications and many more. A cyclic prefix (CP) greater than the maximum length of channel delay spread is appended in OFDM symbol to avoid intersymbol interference (ISI). These signals can be easily demodulated with the help of a simple one-tap equalizer. However, one of the major disadvantages of OFDM system is the sensitivity to carrier frequency offset (CFO) error, which breaks the orthogonality among subcarriers and results in intercarrier interference (ICI), degrading the system performance severely. In this chapter, we study the effect of using phase rotation or compensation with data repetition-based ICI cancellation schemes. This study focuses on the ICI cancellation scheme and its effect on OFDM communication system in AWGN and fading channels, assuming that the synchronization, including phase, frequency and timing has been done by using repeated preamble sequence, but the ICI may still exist due to frequency offset estimation error or unexpected Doppler velocity. It is observed that the improvement obtained in the performance of the ICI cancellation scheme is same, when the information of frequency offset required for phase rotation or compensation is used either at the transmitter or at the receiver.

15.1 INTRODUCTION

The next-generation wireless technology is expected to play a vital role in the field of information and communication technology. It is suggested that the 5G New Radio (NR) technology will operate in a sub-1- to 100-GHz spectrum. It is also assumed that this technology will support diverse services like Internet of things (IoT), Multimedia Broadcast Network, and enhanced mobile broadband. The NR technology has been currently standardized in the 3rd Generation Partnership Project (3GPP). Narrowband IoT (NB-IoT) is introduced in 3GPP release 13. The orthogonal frequency division multiplexing (OFDM) will be used as a modulation in NB-IoT in its downlink (DL) transmission, while single-carrier frequency division multiplexing will access in uplink (UL) transmission.[1] As suggested in 3GPP release 15, the NR systems will adopt OFDM as a modulation technique for both UL and DL transmissions.[2] In future, OFDM will be used in millimeter-wave spectrum in IoT-based 5G networks.

Due to several advantages of OFDM, it may be used in 5G NR techniques, IoTs, and many more applications. A few of the advantages of OFDM include its robustness to frequency-selective fading channels and impulse noises. The use of cyclic prefix (CP) completely eliminates intersymbol interference in OFDM systems. The implementation of OFDM is very simple because of the fast Fourier transform (FFT) algorithms. OFDM systems have a lot of advantages but suffer from several problems like high value of out-of-band radiation, large value of peak-to-average power ratio, and sensitivity to time and frequency synchronization errors. Time and frequency synchronization errors break the orthogonality among the subcarriers of OFDM and result in intercarrier interference (ICI). Due to this system, performance degrades severely. The drawbacks of OFDM may be eliminated to make it a suitable modulation scheme for IoT devices.

To achieve better performance of OFDM systems, an ICI cancellation scheme is presented in this chapter.

15.2 OFDM SYSTEM MODEL

A conventional OFDM system model is presented in Figure 15.1. Here, N subcarriers are used in an OFDM symbol of time period T (T = NTs), where Ts denotes the duration of input data bit/symbol. In the mth OFDM

Orthogonal Frequency Division Multiplexing for IoT

symbol period, the inverse fast Fourier transform (IFFT) processor modulates a group of N data symbols (a0,a1,...,aN–1) onto the N subcarriers.

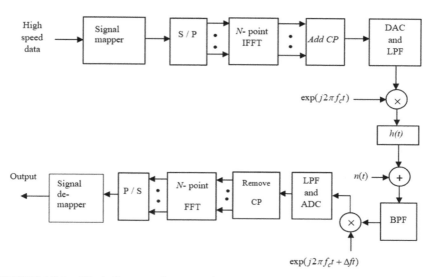

FIGURE 15.1 Block diagram of a conventional OFDM communication system.

At the transmitter of the OFDM system, the baseband-transmitted signal sample of mth OFDM symbol is given as

$$x_n = \frac{1}{N} \sum_{q=0}^{N-1} a_q \exp\left(\frac{j2\pi nq}{N}\right); \quad n = 0,1,\ldots,N-1 \quad (15.1)$$

where $a_q \in \Upsilon$ represents the mapped data symbol on the qth subcarrier with Υ as the M-ary alphabet. The extended symbol after the addition of CP is expressed as

$$x_n^{cp} = \begin{cases} x_{N-N_{cp}+n}; & n \leq N_{cp} \\ x_{n-N_{cp}}; & N_{cp} < n \leq N + N_{cp} \end{cases}; \quad n = 0,1,\ldots,N-1 \quad (15.2)$$

where the length of CP is N_{cp}.

The receiver demodulates the OFDM signal. It is assumed that the transmitter and receiver frequencies are not same. The mismatch in these frequencies may be due to motion between transmitter or receiver or both.

If the mismatch in frequencies is Δf, the samples of received signal, taken at optimum instants after removing CP, suffering from the noise are given as

$$y_n = x_n \exp\left(\frac{j2\pi n \Delta f T}{N}\right) + w_n$$

$$= x_n \exp\left(\frac{j2\pi n \varepsilon}{N}\right) + w_n; \quad n = 0, 1, \ldots, N-1 \tag{15.3}$$

where ε denotes the normalized value of the carrier frequency offset (CFO), and w_n noise samples.

FFT block processes this signal and produces the output as

$$Y_r = \sum_{n=0}^{N-1} y_n \exp\left(\frac{-j2\pi nr}{N}\right) = \sum_{n=0}^{N-1}\left(x_n \exp\left(\frac{j2\pi n \Delta f T}{N}\right) + w_n\right)\exp\left(\frac{-j2\pi nr}{N}\right)$$

$$= \frac{1}{N}\sum_{q=0}^{N-1} a_q \sum_{n=0}^{N-1} \exp\left(\frac{-j2\pi n(r-q-\varepsilon)}{N}\right) + \sum_{n=0}^{N-1} w_n \exp\left(\frac{-j2\pi nr}{N}\right) \tag{15.4}$$

Using the property of geometric series, eq 15.4 can be described by the following expression[3]

$$Y_r = \sum_{q=0}^{N-1} a_q \left(\frac{\sin \pi(r-q-\varepsilon)}{N\sin((\pi(r-q-\varepsilon))/N)}\right) \exp\left(\frac{j\pi(1-N)(r-q-\varepsilon)}{N}\right) + \sum_{n=0}^{N-1} w_n \exp\left(\frac{-j2\pi nr}{N}\right) \tag{15.5}$$

Let $W_r = \sum_{n=0}^{N-1} w_n \exp((-j2\pi nr)/N)$ be the discrete Fourier transform of w_n and

$$c(v) = \frac{1}{N}\left(\frac{\sin(\pi v)}{\sin(\pi v/N)}\right)\exp\left(\frac{j\pi v(1-N)}{N}\right) \tag{15.6}$$

where $v = (r-q-\varepsilon)$.

Equation 15.5 is simplified as

$$Y_r = \sum_{q=0}^{N-1} a_q c(r-q-\varepsilon) + W_r$$

$$= a_r c(-\varepsilon) + \sum_{q=0, q\neq r}^{N-1} a_q c(r-q-\varepsilon) + W_r; \quad r = 0, 1, \ldots, N-1 \tag{15.7}$$

The required signal, which is attenuated in amplitude by a factor $c(-\varepsilon)$, is represented by the first part of eq 15.7, whereas the second part of eq 15.7 represents ICI. The term $c(r-q-\varepsilon)$ denotes weighting coefficients that are responsible for ICI.[3]

It can be observed from eq 15.7 that for $\varepsilon = 0$, the $Y_r = a_r$, that is, the second term in eq 15.7 is zero, and for $\varepsilon \neq 0$, the transmitted data symbol a_r is attenuated in amplitude and affected by other $N-1$ subcarriers.

For a given value of ε, the weighting coefficients of eq 15.6 with modified notations[3] are given as

$$c(q-r) = \frac{1}{N}\left(\frac{\sin(\pi(q-r+\varepsilon))}{\sin((\pi(q-r+\varepsilon))/N)}\right)\exp\left(\frac{j\pi(q-r+\varepsilon)(N-1)}{N}\right) \quad (15.8)$$

The weighting coefficients computed from eq 15.8 are shown in Figure 15.2, for $N = 32$ and $\varepsilon = 0.05$ and $\varepsilon = 0.03$, respectively. There are N different values of these coefficients as $c(0), c(1),..., c(N-1)$.[3] It can also be observed that there is no abrupt change in the values of these coefficients when N changes from -31 to 31, except in the region $\{-1 \leq N \leq 1\}$.

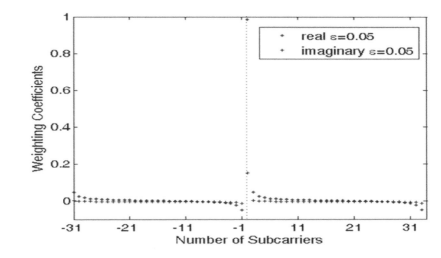

FIGURE 15.2 Weighting coefficients for $N = 32$ and $\varepsilon = 0.05$.

The value of the desired weighting coefficient $c(0)$ decreases with the increase in CFO as revealed by these figures.

The carrier-to-interference ratio (CIR) for standard OFDM system is derived from eq 15.7 as

$$CIR_{std} = \frac{|a_0 c(-\varepsilon)|^2}{\sum_{q=1}^{N-1} |a_q c(-q-\varepsilon)|^2} \quad (15.9)$$

Equation 15.9 can be further simplified as

$$CIR_{std} = \frac{|c(-\varepsilon)|^2}{\sum_{q=1}^{N-1} |c(-q-\varepsilon)|^2} \quad (15.10)$$

Just by simply increasing the transmitted power, ICI cannot be reduced/eliminated because increase in transmission power increases the amplitude of desired as well as ICI terms. Therefore, to reduce the effect of ICI, some strategy be adopted for OFDM systems to make them insensitive for CFO.

15.3 ICI CANCELLATION SCHEMES

In this section, a very simple scheme known as self-ICI cancellation—[3,4] is presented. The scheme reduces the effect of CFO just by rearranging the input data symbols before modulating a subcarrier. The other variants of self-ICI cancellation scheme are also discussed briefly in the proceeding sections.

15.3.1 SELF-ICI CANCELLATION SCHEME

In this scheme, the input data symbol pair $(a_r, -a_r)$ modulates the adjacent subcarriers rth and $(r+1)$th. In this way, the interference generated by a subcarrier cancels automatically with the interference generated by the adjacent subcarrier.[3,4]

The received output (output of the FFT block) is given as

$$Y_r = \sum_{q=0}^{N-1} a_q c(r-q-\varepsilon) + W_r \quad (15.11)$$

and

$$Y_{r+1} = \sum_{q=0}^{N-1} a_q c(r+1-q-\varepsilon) + W_{r+1} \quad (15.12)$$

since the same symbol is used to modulate rth and $(r+1)$th carrier. The received signals corresponding to these subcarriers are written as

$$Y_r = \sum_{\substack{q=0 \\ q=even}}^{N-2} a_q \{c(r-q-\varepsilon) - c(r-q-1-\varepsilon)\} + W_r \quad (15.13)$$

and

$$Y_{r+1} = \sum_{\substack{q=0 \\ q=even}}^{N-2} a_q \{c(r+1-q-\varepsilon) - c(r-q-\varepsilon)\} + W_{r+1} \quad (15.14)$$

To improve the CIR and reduce the ICI, the received symbols Y_r, Y_{r+1}, must be subtracted in pairs.[3,4] Thus, the desired signal corresponding to the rth subcarrier becomes

$$\hat{Y}_r = Y_r - Y_{r+1}$$
$$= \sum_{\substack{q=0 \\ q=even}}^{N-2} a_q \{(2c(r-q-\varepsilon) - c(r-q-1-\varepsilon) - c(r-q+1-\varepsilon))\} + W_r' \quad (15.15)$$

or

$$\hat{Y}_r = a_r \{2c(-\varepsilon) - c(-1-\varepsilon) + c(1-\varepsilon)\}$$
$$+ \sum_{\substack{q=0, q \neq r \\ q=even}}^{N-2} a_q \{2c(r-q-\varepsilon) - c(r-q-1-\varepsilon) + c(r-q+1-\varepsilon)\} + W_r' \quad (15.16)$$

where $W_r' = W_r - W_{r+1}$.

The CIR for this scheme can be obtained from eq 15.16 as

$$CIR_{self} = \frac{|2c(-\varepsilon) - c(-1-\varepsilon) + c(1-\varepsilon)|^2}{\sum_{\substack{q=2 \\ q=even}}^{N-2} |2c(-q-\varepsilon) - c(-q-1-\varepsilon) + c(-q+1-\varepsilon)|^2} \quad (15.17)$$

This scheme cancels the effect of ICI but it does not remove the CPE. Another scheme known as adjacent conjugate symbol repetition (ACSR) cancels the interference by modulating two adjacent subcarriers (r, $r + 1$) by input symbol pair (a_r, $-a_r^*$) presented in Ref. [5]. The operator $(\cdot)^*$ represents the complex conjugate. In this scheme, the effect of ICI and

CPE reduces greatly. The scheme works effectively only for small values of CFOs.

15.3.2 CONJUGATE CANCELLATION (CC) SCHEME

In this scheme, for the transmission of OFDM signals, two paths of signal processing are used. The standard OFDM signal is transmitted in the first path, whereas in the second path the conjugate signal of the first path is transmitted. The received signal is expressed in two parts as

$$y_n^{(1)} = x_n \exp\left(\frac{j2\pi\varepsilon n}{N}\right) + w_n^{(1)} \qquad (15.18)$$

and

$$y_n^{(2)} = x_n^* \exp\left(\frac{j2\pi\varepsilon n}{N}\right) + w_n^{(2)} \qquad (15.19)$$

where x_n^* is complex conjugate of x_n, $w_n^{(i)}$; $i = 1, 2$ represent noise samples.

The signals $y_n^{(1)}$ and $\left(y_n^{(2)}\right)^*$ are passed through the DFT blocks to get the frequency domain signal samples. For getting the desired signal samples, average of $y_n^{(1)}$ and $y_n^{(2)}$ is computed as in Ref [6].

The CIR for the present scheme under no noise condition is given as

$$CIR_{CC} = \frac{|c(-\varepsilon) + c(\varepsilon)|^2}{\sum_{q=1}^{N-1} |c(q+\varepsilon) + c(q-\varepsilon)|^2} \qquad (15.20)$$

The improvement in the CIR and bit error rate (BER) performances of the present scheme as compared to standard, ICI self-cancellation, and ACSR schemes is observed. However, this improvement can be noticed for low-frequency offset values only. In the high-frequency offset region, CPE correction and ICI reduction are not obtained from this scheme. Thus, the improvement in the CIR for the present scheme cannot be observed.[7]

15.3.3 PRCC SCHEME

This scheme is modified version of the scheme discussed in Section 15.3.2. In the PRCC scheme, the first-path signal of conjugate cancellation (CC) scheme is modified by a phase rotation ϕ and conjugate of

Orthogonal Frequency Division Multiplexing for IoT

the second-path signal of CC scheme is modified by an artificial phase rotation of $-\phi$. The ICI induced in the system is effectively cancelled at the receiver by the use of this phase rotation.

The received signal in the presence of CFO corresponding to the two transmission paths is given as

$$y_n^{(1)} = x_n e^{j\phi} \exp\left(\frac{j2\pi\varepsilon n}{N}\right) + w_n^{(1)}; \quad n = 0,1,\ldots,N-1 \quad (15.21)$$

and

$$y_n^{(2)} = x_n^* e^{j\phi} \exp\left(\frac{j2\pi\varepsilon n}{N}\right) + w_n^{(2)} \quad (15.22)$$

The FFT of $y_n^{(1)}$ and $\left(y_n^{(2)}\right)^*$ is taken to obtain the following frequency domain signals.

$$Y_r^{(1)} = FFT\{y_n^{(1)}\} = \sum_{q=0}^{N-1} a_q \{e^{j\phi} c(r-q-\varepsilon)\} + W_r^{(1)} \quad (15.23)$$

$$Y_r^{(2)} = FFT\{(y_n^{(2)})^*\} = \sum_{q=0}^{N-1} a_q \{e^{-j\phi} c(r-q+\varepsilon)\} + W_r^{(2)} \quad (15.24)$$

where $W_r^{(i)} = FFT\{w_r^{(i)}\}$; $i = 1,2$. The final desired signal corresponding to the rth subcarrier is obtained as

$$\begin{aligned} Z_r &= \frac{1}{2}\{Y_r^{(1)} + Y_r^{(2)}\} \\ &= \frac{1}{2}\left\{\sum_{q=0}^{N-1} a_q \left(e^{j\phi} c(r-q-\varepsilon) + e^{-j\phi} c(r-q+\varepsilon)\right) + \left(W_r^{(1)} + W_r^{(2)}\right)\right\} \end{aligned} \quad (15.25)$$

The CIR of the PRCC scheme is expressed from eq 15.25 as

$$CIR_{PRCC} = \frac{|a_0|^2 \left|e^{j\phi} c(r-\varepsilon) + e^{-j\phi} c(r+\varepsilon)\right|^2}{\sum_{q=1}^{N-1} |a_q|^2 \left|e^{j\phi} c(r-q-\varepsilon) + e^{-j\phi} c(r-q+\varepsilon)\right|^2} \quad (15.26)$$

For equal symbol energies, eq 15.26 can be simplified as

$$CIR_{PRCC} = \frac{\left|e^{j\phi}c(-\varepsilon)+e^{-j\phi}c(\varepsilon)\right|^2}{\sum_{q=1}^{N-1}\left|e^{j\phi}c(-q-\varepsilon)+e^{-j\phi}c(-q+\varepsilon)\right|^2} \quad (15.27)$$

The optimum value of phase rotation ϕ is derived in Ref. [10] as

$$\phi_{opt} = -\pi\varepsilon\frac{N-1}{N} \quad (15.28)$$

The use of eq 15.28 suggests that the information about CFO is required at the transmitter for obtaining optimum phase rotation. Therefore, the CFO estimation is carried out at the receiver and the information about it is sent back to the transmitter, where optimum value of ϕ is computed.

The PRCC scheme improves the BER performance, but for higher values of ε, its CIR degrades and also increases the transmitter complexity.

15.4 PCSC SCHEME (PROPOSED SCHEME)

In the standard OFDM system, the received signal samples are attenuated by a common term $c(-\varepsilon)$ as given in eq 15.7. This common term can be approximated as $c(-\varepsilon) \simeq \exp(j\phi_c)$ for $\varepsilon < 1$, where $\phi_c \simeq \pi\varepsilon((1-N)/N)$ is termed as CPE.[5] The CPE depends on ε and can be compensated either at the transmitter or at the receiver after estimation of ε. The estimation of CFO can be done at the receiver with the help of any existing frequency offset estimation methods.[8,9] After estimation of CFO ε, its effect can be cancelled either at the receiver itself or by sending it back to the transmitter. It is observed through simulations that the BER performance of the present scheme remains the same if the phase is compensated either at the transmitter or at the receiver. The diagram of the proposed method is shown in Figure 15.3. In the proposed scheme, also termed PCSC (phase compensated self-cancellation) scheme, the information about phase ϕ_c is estimated and compensated at the receiver.

Orthogonal Frequency Division Multiplexing for IoT

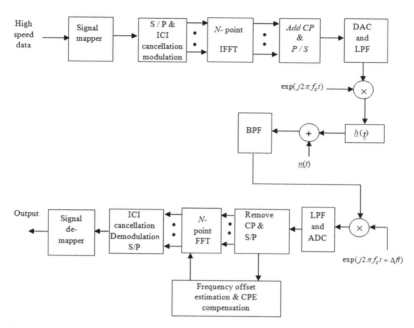

FIGURE 15.3 Block diagram of proposed PCSC OFDM communication system.

The CPE-compensated discrete-time samples of received OFDM signal are given as

$$y_n = \left\{ x_n \exp\left(\frac{j2\pi\varepsilon n}{N}\right) + w_n \right\} \exp(-j\phi_c); \quad n = 0, 1, \ldots, N-1 \quad (15.29)$$

where $\{x_n\}$ is the inverse discrete Fourier transform (IDFT) of $\{a_q\}$, that is, $x_n = IDFT\{a_q\}$ and w_n are AWGN samples. The data symbols $\{a_q\}_{q=0}^{N-1}$ are the transmitted symbols so constrained that $\{a_1 = -a_0, a_3 = -a_2, \ldots, a_{N-1} = -a_{N-2}\}$ as in the case of ICI self-cancellation scheme.[3]

After compensating for CFO, the signal samples on subcarrier r and (r + 1) at the output of DFT block are given as

$$Y_r = \sum_{\substack{q=0 \\ q=even}}^{N-2} \{a_q \left(c'(r-q-\varepsilon) - c'(r-q-1-\varepsilon) \right)\} + W_r \quad (15.30)$$

and

$$Y_{r+1} = \sum_{\substack{q=0 \\ q=even}}^{N-2} \{a_q(c'(r+1-q-\varepsilon) - c'(r-q-\varepsilon))\} + W_{r+1} \quad (15.31)$$

where $c'(r-q-\varepsilon) = c(r-q-\varepsilon)\exp(-j\phi_c); W_r$; W_r and W_{r+1} are the DFT of $w_n \exp(-j\phi_c)$ and $w_{n+1}\exp(-j\phi_c)$, respectively.

Since same data symbols are transmitted on the adjacent subcarriers, the resultant samples of the received signal are obtained from eqs 15.30 and 15.31 as

$$Y'_r = Y_r - Y_{r+1}$$
$$= \sum_{\substack{q=0 \\ q=even}}^{N-2} \{a_q(2c'(r-q-\varepsilon) - c'(r-q-1-\varepsilon) - c'(r-q+1-\varepsilon))\} + W'_r \quad (15.32)$$

where $W'_r = W_r - W_{r+1}$.

The CIR for the proposed scheme can be obtained from eq 15.32 for equal symbol energies as

$$CIR_{proposed} = \frac{|2c'(-\varepsilon) - c'(-1-\varepsilon) - c'(1-\varepsilon)|^2}{\sum_{\substack{q=0 \\ q=even}}^{N-2} |2c'(-q-\varepsilon) - c'(-q-1-\varepsilon) - c'(-q+1-\varepsilon)|^2} \quad (15.33)$$

The expressions in eqs 15.23 and 15.24 and eqs 15.30 and 15.31 have been obtained for AWGN channel. The expressions in the case of fading channels can also be obtained if the channel gains are taken into account. A frequency domain equalizer under the known channel condition as considered in Ref. [7] can be utilized for compensation of the channel gains, and the expressions in eqs 15.30 and 15.31 can be derived in the same way.

In the PCSC scheme, the artificial phase rotation at the transmitter does not require and therefore, the transmitter's complexity is less as compared to PRCC scheme. However, the receiver's complexity of both the schemes is the same. Also, to supply the required phase rotation to transmitter, no feedback channel is required in this scheme.

15.5 SIMULATION RESULTS

To evaluate the performance of the PCSC scheme, some other methods, namely, ICI self-cancellation, CC, and the PRCC schemes, are also simulated under noise and multipath fading environments. The OFDM system with 64 subcarriers, PSK/QAM data symbols, and CFO chosen as 0.05, 0.1, and 0.2 is considered for the study. The transmission carrier of 2.4 GHz is considered in the analysis. Each OFDM symbol is appended by 16 samples of CP.

The BER performances of various schemes for coded and uncoded OFDM systems for different values of CFOs are shown in Figures 15.4–15.6 for AWGN channel. From Figure 15.4, it is clear that the BER of the present scheme is similar to that of self-ICI cancellation and PRCC scheme at $\varepsilon = 0.05$. Furthermore, if CFO is large ($\varepsilon \geq 0.2$), the improvement in the BER performance of the proposed scheme is as compared to ICI self-cancellation and PRCC schemes.

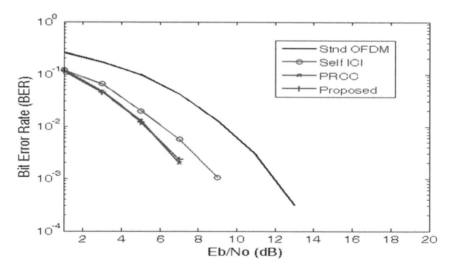

FIGURE 15.4 BER plot for various uncoded schemes under AWGN channel at constant $\varepsilon = 0.05$.

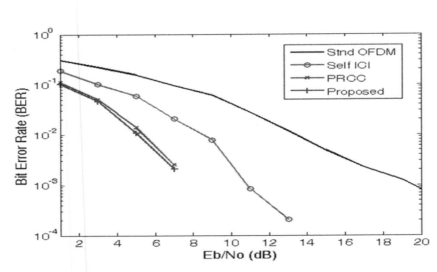

FIGURE 15.5 BER plot for various uncoded schemes under AWGN channel at constant $\varepsilon = 0.1$.

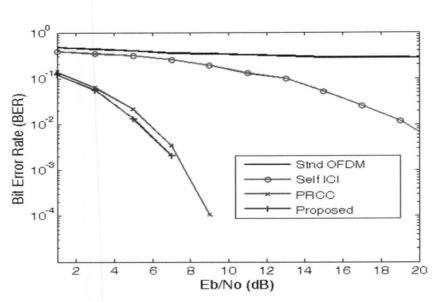

FIGURE 15.6 BER plot for various uncoded schemes under AWGN channel at constant $\varepsilon = 0.2$.

Orthogonal Frequency Division Multiplexing for IoT

Figures 15.7–15.9 show the BER performance of a (7, 4) linear block–coded OFDM systems at different values of CFO under AWGN channel. An improved BER performance is also obtained for coded schemes.

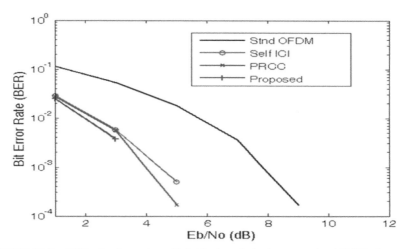

FIGURE 15.7 BER plot for various (7,4) block-coded schemes under AWGN channel at constant $\varepsilon = 0.05$.

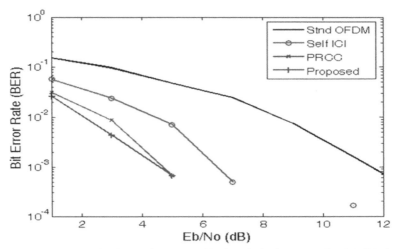

FIGURE 15.8 BER plot for various (7, 4) block-coded schemes under AWGN channel at constant $\varepsilon = 0.1$.

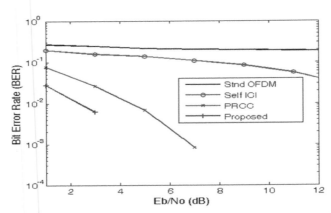

FIGURE 15.9 BER plot for various (7, 4) block-coded schemes under AWGN channel at constant $\varepsilon = 0.2$.

Next, the BER performances of these systems are evaluated through simulations under multipath fading environment. The wireless fading channel considered in simulation is the six-path Typical Urban channel model described in Refs. [10,11]. The BER performance is evaluated under the condition of perfect channel state information as considered in Ref. [10]. As shown in Figures 15.10–15.12, the BER of present scheme is similar to PRCC scheme for small frequency offset ($\varepsilon = 0.05$), while at large CFO ($\varepsilon \geq 0.2$), the BER of the proposed scheme is much better than that of ICI self-cancellation and PRCC schemes.

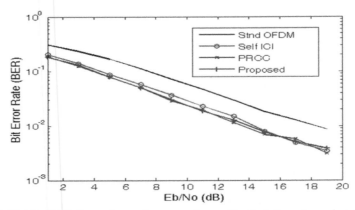

FIGURE 15.10 BER for various schemes under multipath fading channel at constant $\varepsilon = 0.05$.

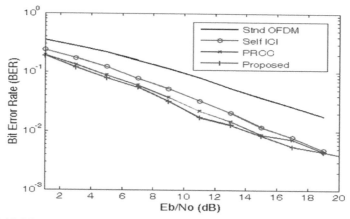

FIGURE 15.11 BER for various schemes under multipath fading channel at constant $\varepsilon = 0.1$.

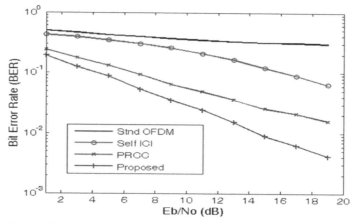

FIGURE 15.12 BER for various schemes under multipath fading channel at constant $\varepsilon = 0.2$.

Figures 15.13–15.15 depict the BER versus E_b/N_o plot for different (7, 4) block-coded schemes under multipath Rayleigh fading channel. The proposed scheme still performs better as compared to other ICI cancellation schemes. It can be observed from these figures that the proposed scheme performs better not only in AWGN channel but also in multipath fading scenarios.

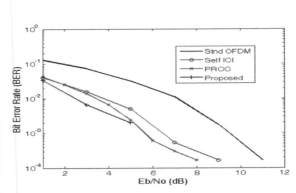

FIGURE 15.13 BER for various (7, 4) block-coded schemes under multipath fading channel at constant $\varepsilon = 0.05$.

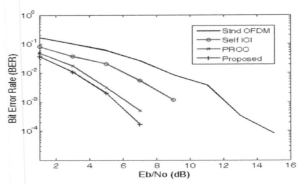

FIGURE 15.14 BER for various (7, 4) block-coded schemes under multipath fading channel at constant $\varepsilon = 0.1$.

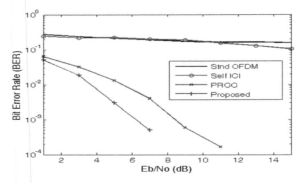

FIGURE 15.15 BER for various (7, 4) block-coded schemes under multipath fading channel at constant $\varepsilon = 0.2$.

The main aim of ICI cancellation is to improve the BER performance of the system. The BER of an OFDM system is affected by three factors—ICI, CPE, and noise, out of which first two are dependent on CFO. When the CFO is small, the values of ICI and CPE are also small and their effect is not critical, as it does not change the decision on the symbol value over a large range of signal-to-noise ratio (SNR) as shown by the scatter plots of Figure 15.16. However, when CFO becomes significant, the BER of the system is affected even at high SNR. As a result, BER of the system increases as may be noticed from Figure 15.17. Thus, the BER of the present scheme is similar to other schemes at small values of CFO although it has slightly smaller CIR in comparison with CC and PRCC schemes (as depicted in Fig. 15.18).

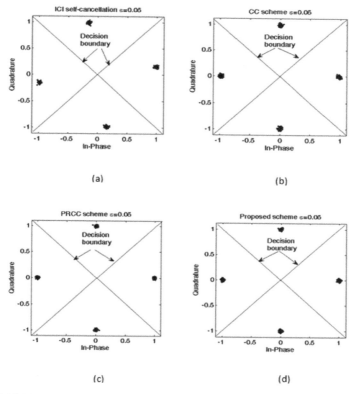

FIGURE 15.16 Scatter plots of received signal for (a) ICI self-cancellation scheme, (b) CC scheme, (c) PRCC scheme, and (d) proposed scheme at SNR = 30 dB and $\varepsilon = 0.05$.

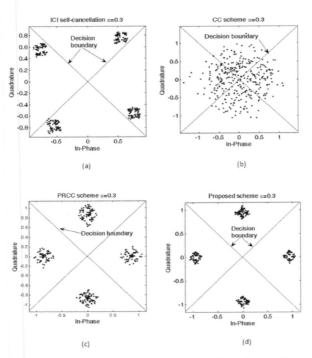

FIGURE 15.17 Scatter plots of received signal for (a) ICI self-cancellation scheme, (b) CC scheme, (c) PRCC scheme, and (d) proposed scheme at SNR = 30 dB and $\varepsilon = 0.3$.

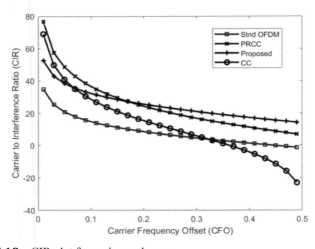

FIGURE 15.18 CIR plot for various schemes.

Orthogonal Frequency Division Multiplexing for IoT

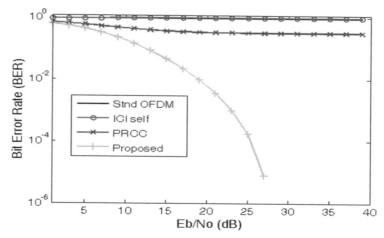

FIGURE 15.19 BER for various schemes under AWGN channel at $\varepsilon = 0.4$, 16-PSK, and for $N = 1024$.

Figure 15.18 depicts CIR comparisons of the normal OFDM system, CC scheme, PRCC scheme, and the present scheme. The CIR of various schemes is plotted under the assumption of exact phase rotation/phase correction (ϕ / ϕ_c) with respect to e and zero-noise condition. From this figure, it is evident that the CIR of the proposed scheme is poorer than the PRCC scheme up to e $\approx$ 0.17. However, its BER performance is comparable with the PRCC scheme and better than the ICI self-cancellation scheme.

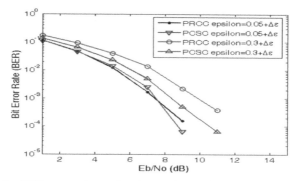

FIGURE 15.20 BER comparison of various schemes for both fixed and time-varying frequency offset for AWGN channel.

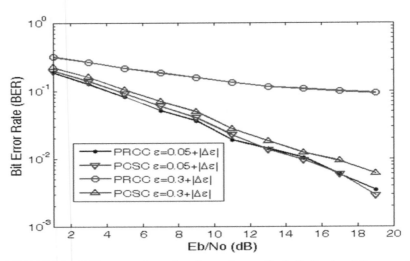

FIGURE 15.21 BER comparison of various schemes for both fixed and time-varying frequency offset for multipath fading channel.

At higher values of CFO, the BER is affected by ICI and CPE due to their large values. For $\varepsilon > 0.17$, the CIR of the present scheme is better than PRCC scheme and this translates into superior BER performance as shown in Table 15.1.

The PRCC and present schemes rely on cancellation of CPE for improvement in their performances. However, the performance of the PCSC scheme is much better than the PRCC scheme because of self-cancellation of ICI along with CPE correction. It is also noted that PRCC and CC schemes have better CIR than the present scheme for low CFO. However, it does not translate into any improvement in BER performance.

TABLE 15.1 CIR and BER for Various Schemes at SNR = 10 dB.

Schemes	$\varepsilon = 0.1$		$\varepsilon = 0.4$	
	CIR (dB)	BER (%)	CIR (dB)	BER (%)
CC	27.9703	0	6.4885	58
PRCC	35.0758	0	9.5292	1
PCSC (proposed)	32.4368	0	17.2529	0.02

It has also been observed through analysis that the proposed scheme performs better than other schemes for higher order signal constellation (i.e., for 16-PSK and higher) and for large values of ε. To observe this behavior, conventional OFDM system, self-ICI cancellation scheme, PRCC scheme, and the present scheme are also simulated for 1024 subcarriers with 16-PSK-modulated input data symbols and $\varepsilon = 0.4$. The BER versus E_b/N_o performance of all these schemes under AWGN channel is plotted in Figure 15.19. It is revealed from Figure 15.19 that the PCSC scheme provides better BER performance than the other mentioned schemes for $N > 64$, $\varepsilon > 0.2$, and for higher order signal constellations.

It is also important to examine the effect of CFO estimation error or time variation in CFO on the performance of the proposed and PRCC schemes. Therefore, the BER performances of PCSC and PRCC schemes are investigated under CFO estimation error. To simulate this scenario, we consider that the fixed value of CFO $\varepsilon = 0.05$ and 0.3 and the error in CFO estimation (or variation in CFO due to sudden change in the speed of a mobile station, i.e., Doppler shift) $\Delta\varepsilon$ are random and uniformly distributed over $\Delta\varepsilon \in (-0.1, 0.1)$.

The BER of the PCSC scheme and PRCC scheme at fixed and time-varying CFOs is plotted in Figure 15.20 for AWGN channel and in Figure 15.21 for a six-path wireless fading channel. From these figures, it is observed that the performance of the PCSC scheme is comparable with that of the PRCC scheme with CFO estimation error when fixed value of CFO is small, that is, $\varepsilon = 0.05$, and it is much better when fixed value of CFO is large, that is, $\varepsilon = 0.3$. Therefore, it may be stated that the PCSC scheme may be suitable for cancellation of ICI under this case also.

15.6 CONCLUSION

The performance of a data repetition–based ICI cancellation scheme, in general, improves significantly when it is combined with the information about the frequency offset. This may be implemented either in the form of phase rotation at the transmitter or, alternatively, as phase compensation at the receiver. Although the performance remains the same in both cases, the phase compensation at receiver is of greater benefit as it obviates the need of feedback path. Further, the combination of phase compensation with the standard ICI self-cancellation scheme[3,4] yields much better performance at high values of frequency offset than when the information about the frequency offset is used with two-path CC scheme. At small values of frequency offset, BER of present scheme is almost the same as that of other schemes. Thus, if OFDM systems are ICI free, they may support a tremendous data rate and may be used in IoT-based 5G networks.

KEYWORDS

- **OFDM**
- **CIR**
- **ICI cancellation**
- **PRCC**
- **PCSC**

REFERENCES

1. Xu, T.; Darwazeh, I. Non-Orthogonal Narrowband Internet of Things: A Design for Saving Bandwidth and Doubling the Number of Connected Devices. *IEEE Internet of Things J.* **2018**, *5*, 2120–2129.
2. Zaidi, A. A.; Baldemair, R.; Moles-Cases, V.; He, N.; Werner, K.; Cedergen, A. OFDM Numerology Design for 5G New Radio to Support IoT, eMBB, AND mbsfn. *IEEE Commun. Standards Mag.* **2018**, *2*, 78–83.
3. Armstrong, J. Analysis of New and Existing Methods of Reducing Intercarrier Interference Due to Carrier Frequency Offset in OFDM. *IEEE Trans. Commun.* **1999**, *73*, 365–369.
4. Zhao, Y.; Häggman, S.-G. Intercarrier Interference Self Cancellation Scheme for OFDM Mobile Communication Systems. *IEEE Trans. Commun.* **2001**, *49*, 1185–1191.
5. Sathananthan, K.; Athaudage, C. R. N.; Qiu, B. In *A Novel ICI Cancellation Scheme to Reduce Both Frequency Offset and IQ Imbalance Effects in OFDM*, Proc. IEEE 9th Int. Symp. Comput. Commun. 2004, 708–713.
6. Yeh, H.-G.; Chang, Y.-K.; Hassibi, B. A Scheme for Cancelling Intercarrier Interference Using Conjugate Transmission in Multicarrier Communication Systems. *IEEE Trans. Wireless Commun.* **2007**, *6*, 3–7.
7. Wang, C.-L.; Huang, Y.-C. Intercarrier Interference Cancellation Using General Phase Rotated Conjugate Transmission for OFDM Systems. *IEEE Trans. Commun.* **2010**, *58*, 812–819.

8. Zeng, X.; Ghrayeb, A. A Blind Carrier Frequency Offset Estimation Scheme for OFDM Systems with Constant Modulus Signaling. *IEEE Trans. Commun.* **2008**, *56*, 1032–1037.
9. Kumar, A.; Pandey, R. A Bandwidth Efficient Method for Cancellation of ICI in OFDM Systems. *Int. J. Electron. Commun. (AEÜ)* **2009**, *63*, 569–575.
10. Pun, Man-On; Morelli, M.; Kuo, C.-C. J. *Multi-carrier Echniques for Broadband Wireless Communications*; Imperial College Press: London, 2007.
11. Li, Y. (G.); Seshadri, N.; Ariyavisitakul, S. Channel Estimation for OFDM Systems with Transmitter Diversity in Mobile Wireless Channels. *IEEE J. Selected Areas Commun.* **1999**, *17*, 461–471.

CHAPTER 16

Fading Channel Capacity of Cognitive Radio Networks

INDU BALA*

School of Electronics and Electrical Engineering, Lovely Professional University, Phagwara, Punjab, India

*E-mail: i.rana80@gmail.com

ABSTRACT

Due to the existing spectrum scarcity issue, it has become extremely difficult to deploy new wireless communication systems/ applications. Recently, cognitive radio technology has evolved very rapidly to address this challenge by allowing unlicensed users to exploit the transmission channel of the licensed user opportunistically whenever primary user is not using it. It not only improves the spectrum utilization but also enhances the spectrum efficiency manifolds. In this chapter, two different transmission power control schemes are assessed for various system parameters by exploiting channel sensing information under received power constraints. These schemes are: (i) Adaptive transmission power scheme and (ii) Adaptive rate and transmission power scheme. It has been demonstrated through simulated results that adaptive transmission power outperforms the second scheme for given bit error rate.

16.1 INTRODUCTION

The introduction of bandwidth-starving IoT applications in our lives has elicited a gigantic bandwidth requirement and it is projected to cultivate more in near upcoming years. The old-fashioned spectrum sharing strategies in use provide licensed primary users (PUs) long-term access to the frequency bands over a topographical area. Therefore, accommodating imminent wireless applications or services in the present scenario has become tremendously difficult. The state-of-the-art report from regulatory bodies has publicized that the whole allocated frequency spectrum is being used intermittently approximately 15–85%. Thus, the conclusion is that the prevailing spectrum paucity problem is primarily due to the bungling practice of spectrum usage rather than the genuine scarcity.[1] Recently the notion of cognitive radio technology has been presented to overcome this spectrum paucity problem. It permits the unlicensed secondary user (SU) to persistently monitor its neighboring RF channel(s) and to acclimatize its system parameters in such a way that both, licensed and unlicensed, users may coexist in the same channel along with PU.[2] Many spectrum sharing modes have been recommended for the dynamic spectrum access, namely, underlay mode, overlay mode, and interweave mode.[3] In underlay mode, the SU is permitted to function concurrently with PU such that the interference experienced by PU is below the threshold bound.[4–6] In the overlay mode of communication, the advanced coding and signal processing techniques are used to facilitate PU by retransmitting its signal.[7–8] Whereas, the interweave mode of communication allows SU to speculatively exploit spectral opportunities through sensing before transmission.[9–12]

In this chapter, a communication system is considered in which unlicensed users access licensed spectrum opportunistically and regulate its system parameters such as transmission power based on channel sensing information under received power constraints. Given the bit error rate (BER), the projected scheme is examined for two different power adaptation schemes and the benefits of using sensing information and CSI are investigated for SU.

The remaining chapter is structured as follows: the system model is presented in Section 16.2. The fading channel capacity of proposed system model is estimated for two different power control schemes in Section 16.3. Finally, the numerical results followed by the conclusion are presented in Sections 16.4 and 16.5, respectively.

16.2 SYSTEM MODEL

We have considered a spectrum sharing system, as shown in Figure 16.1, in which SU is permitted to share the licensed spectrum of PU such that

Fading Channel Capacity of Cognitive Radio Networks

no or minimum interference may be experienced by incumbent receiver. To regulate SU transmission, peak power and average received power constraints have been imposed on PU. If $X_s[n]$ is a SU signal, received SU signal will be given by

$$Y_s[n] = x_s[n]\sqrt{\gamma_s[n]} + z_s[n] \tag{16.1}$$

where γ_s represents the channel power gain between SU transmitter (STx) and SU receiver (SRx) and $z_s[n]$ represents Gaussian noise with variance $N_0 B$. The term N_0 and B represents noise and bandwidth, respectively.

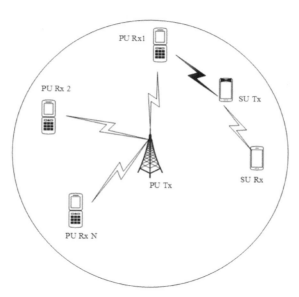

FIGURE 16.1 Proposed spectrum sharing scenario.

We further define $\sqrt{\gamma_p}$ as a channel gain between PU transmitter (PTx) and PU receiver (PRx), $\sqrt{\gamma_{sp}}$ between STx and PRx and $\sqrt{\gamma_m}$ between PTx and STx. The channel gains are the function of actual physical distance between transmitters and receivers, for example, d_{sp}^{-2} for γ_{sp} and d_m^{-2} for γ_m. The STx acclimates its transmission power using sensing statistics without disrupting the received power constraints when PU is active in a channel. These received power constraints considered for communication scenario presented in this chapter are

$$E_{\gamma_s,\gamma_{sp},\xi}\left\{T(\gamma_s,\gamma_{sp},\xi)\gamma_{sp}\right\}_{PU\_On} \leq T_{Avg} \quad (16.2)$$

$$\left\{T(\gamma_s,\gamma_{sp},\xi)\gamma_{sp}\right\}_{PU\_On} \leq T_{Peak}; \quad \forall \ \gamma_s,\gamma_{sp} \ \& \ \xi \quad (16.3)$$

where eqs 16.2 and 16.3 represent average power received and peak power received at PRx, respectively. $T(\gamma_s,\gamma_{sp},\xi)$ symbolizes SU transmission power, and $E_{\gamma_s,\gamma_{sp},\xi}[.]$ symbolizes expectation operator over γ_s, γ_{sp}, and ξ.

16.3 FADING CAPACITY

Under noisy channel condition, fading capacity is used as a performance pointer for the applications in which reliability of data is more important than delay. It represents the maximum possible data rate with random small probability of error. We have obtained the closed-form expressions for the channel capacity of two different transmission schemes. These schemes are adaptive transmission power (ATP) scheme and adaptive rate and transmission power (ARTP) scheme, under received power constraints for given bit error. The approach suggested in Refs. [13,14] for calculating fading channel capacity is used here to achieve optimum power control such that eqs 16.2 and 16.3 are also satisfied. Based on this, the optimization problem may be formulated as

$$\frac{C_{er}}{B} = \max_{T(\gamma_s,\gamma_{sp},\xi)} \left\{ \log_2 \left(1 + \frac{T(\gamma_s,\gamma_{sp},\xi)\gamma_s}{N_0 B} \right) \right\} \quad (16.4)$$

subject to constraints eqs 16.2 and 16.3.

$$\frac{C_{er}}{B} = A\left[-B + C\left\{\log_2(D*E)\right\}\right] \quad (16.5)$$

$$A = \frac{1}{\sqrt{4\pi}}\left[\Gamma\left(0.5, \frac{(T_1(\gamma_u(\xi))-\mu_{on})^2}{2\delta_{on}^2}\right) - \Gamma\left(0.5, \frac{(T_2(\gamma_u(\xi))-\mu_{on})^2}{2\delta_{on}^2}\right)\right] \quad (16.6)$$

$$B = \log_2\left(1 - \frac{T_{Peak}\lambda_1}{N_0 B \lambda_1 + \gamma_u(\xi)}\right) \quad (16.7)$$

$$C = \frac{T_{Peak}}{T_{Peak} + N_0 B}; \quad D = \frac{T_{Peak}\lambda_1}{N_0 B \gamma_u(\xi)} \quad (16.8)$$

$$E = \left(N_0 B + \frac{\gamma_u(\xi)}{\lambda_1} - T_{Peak} \right) \qquad (16.9)$$

Given the values of γ_s, γ_{sp}, and ξ using M-QAM modulation scheme with

$$BER(\gamma_s, \gamma_{sp}, \xi) \leq 0.2 \exp\left(\frac{-1.5}{M-1} \frac{T(\gamma_s, \gamma_{sp}, \xi) \gamma_s}{N_0 B} \right) \qquad (16.10)$$

and

$$C = \frac{-1.5}{\ln(5\ BER)} \leq 1 \qquad (16.11)$$

The fading channel capacity for the system under study becomes the solution of following optimization problem

$$\frac{C_{er}}{B} = \max\left\{ E_{\gamma_s, \gamma_{sp}, \xi} \left[\log_2\left(1 + C \frac{T(\gamma_s, \gamma_{sp}, \xi) \gamma_s}{N_0 B} \right) \right] \right\} \qquad (16.12)$$

subject to eqs 16.2 and 16.3. The fading channel capacity for ARTP scheme with M-QAM scheme will be given by

$$\frac{C_{er}}{B} = A\left[-B + C\{\log_2(D*E)\} \right] \qquad (16.13)$$

$$A = \frac{1}{\sqrt{4\pi}} \left[\Gamma\left(0.5, \frac{(T_1(\gamma_u(\xi)) - \mu_{on})^2}{2\delta_{on}^2} \right) - \Gamma\left(0.5, \frac{(T_2(\gamma_u(\xi)) - \mu_{on})^2}{2\delta_{on}^2} \right) \right] \qquad (16.14)$$

$$B = -\log_2\left(1 - \frac{CT_{Peak}\lambda_1}{N_0 B \lambda_1 + C\gamma_u(\xi)} \right) \qquad (16.15)$$

$$C = \frac{CT_{Peak}}{CT_{Peak} + N_0 B}; \quad D = \frac{CT_{Peak}\lambda_1}{N_0 B\ C\gamma_u(\xi)} \qquad (16.16)$$

$$E = \left(N_0 B + \frac{C\gamma_u(\xi)}{\lambda_1} - CT_{Peak} \right) \qquad (16.17)$$

16.4 NUMERICAL RESULTS

For the communication scenario presented in this chapter, the simulated results are presented in this section to validate theoretical results. Figure 16.2 shows the SU transmission profile under received power constraints.

The value $\gamma_u(\xi) > 1$ exemplifies the likelihood of PU to be sedentary in a common channel, $\gamma_u(\xi) < 1$ exemplifies likelihood of PU to be active, and $\gamma_u(\xi) = 1$ exemplifies a situation of adjusting SU transmission power without sensing results.

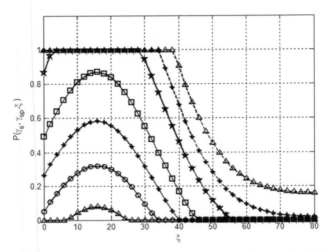

FIGURE 16.2 Instantaneous transmission power versus sensing metric ξ ($P_{peak} \cong 0$ dB).

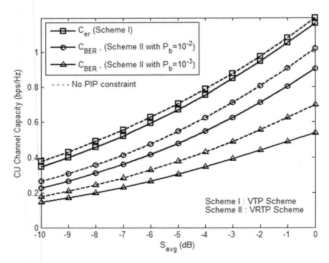

FIGURE 16.3 Fading channel capacity for ARTP scheme with $\rho = 1.7$.

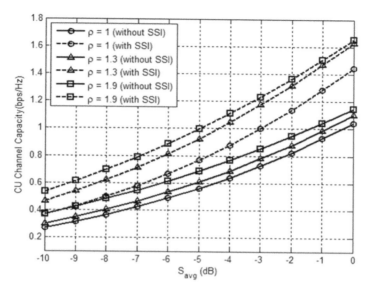

FIGURE 16.4 Comparison for channel capacities of various transmission schemes for different values of ρ.

It is demonstrated that SU adapts its transmission power based on the sensing results in a shared channel. When PU is absent from channel, SU transmits at high power level and when PU is present in a channel, SU reduces its transmission power. The SU fading channel capacity for ARTP scheme using M-QAM as a function of P_{Avg} for different BER is shown in Figure 16.3. The scheme decreases channel capacity drastically irrespective to the sensing results and received power constraints due to parameter C.

Figure 16.4 demonstrates the benefit of using sensing results. It has been witnessed that sensing information helps SU to achieve high fading capacity.

16.5 CONCLUSION

The fading channel capacity for SU is investigated in this chapter for peak and average received power constraints at PRx. The closed-form expression for fading capacity has been derived for two power control schemes,

ATP control scheme and ARTP control schemes. It has been demonstrated with the help of simulation results that the sensing information helps SU to achieve high channel capacity under the said constraints at the PRx. Moreover, it has also been observed that ATP scheme is able to achieve higher capacity than ARTP scheme for given BER.

KEYWORDS

- **cognitive radio**
- **throughput**
- **bit error rate**
- **transmission power control**

REFERENCES

1. Akyildiz, I. F.; Lee, W. Y.; Vuran, M. C.; Mohanty, S.. NeXt Generation/Dynamic Spectrum Access/Cognitive Radio Wireless Networks: A Survey. *Comput. Netw.* **2006**, *50*; 2127–2159.
2. Bala, I.; Bhamrah, M. S.; Singh, G. Analytical Modelling of Ad Hoc Cognitive Radio Environment for Optimum Power Control. *Int. J. Comput. Appl.* **2014**, *92*, 19–22.
3. Sethi, R.; Bala, I. Throughput Enhancement of Cognitive Radio Networks Through Improved Frame Structure. *Int. J. Comput. Appl.* **2015**, *109*, 40–43.
4. Gastpar, M. On Capacity under Receive and Spatial Spectrum Sharing Constraints. *IEEE Trans. Inf. Theor.* **2007**, *53*, 471–487.
5. Ghasemi, A.; Sousa, E. S. Fundamental Limits of Spectrum-Sharing in Fading Environments. *IEEE Trans. Wireless Commun.* **2007**, *6*, 649–658.
6. Suraweera, H.; Smith, P.; Shafi, M. Capacity Limits and Performance Analysis of Cognitive Radio with Imperfect Channel Knowledge. *IEEE Trans. Veh. Technol.* **2010**, *59*, 1811–1822.
7. Jovicic, A.; Viswanath, P. Cognitive Radio: An Information-Theoretic Perspective. *IEEE Trans. Inf. Theor.* **2009**, *55*, 3945–3958.
8. Rana, V.; Jain, N.; Bala, I.; Bhamrah, M. S.; Singh, G. Resource Allocation Models for Cognitive Radio Networks: A Study. *Int. J. Comput. Appl.* **2014**, *91*, 51–55.
9. Asghari, V.; Aissa, S. Adaptive Rate and Power Transmission in Spectrum-Sharing Systems. *IEEE Trans. Wireless Commun.* **2010**, *9*, 3272–3280.
10. Bala, I.; Bhamrah, M. S.; Singh, G. Capacity in Fading Environment Based on Soft Sensing Information under Spectrum Sharing Constraints. *Wireless Netw.* **2017**, *23*, 519–531.

11. Goldsmith, A.; Chua, S. G. Variable Rate Variable Power MQAM for Fading Channels. *IEEE Trans. Commun.* **1997,** *45*,1218–1230.
12. Bala, I.; Bhamrah, M. S.; Singh, G. Rate and Power Optimization under Received-Power Constraints for Opportunistic Spectrum-Sharing Communication. *Wireless Pers. Commun.* **2017,** *96*, 5667–5685.
13. Webb, W. T.; Steele, R. Variable Rate QAM for Mobile Radio. *IEEE Trans. Commun.* **1995,** *43*, 2223–2230.
14. Bala, I.; Bhamrah, M. S.; Singh, G. Investigation on Outage Capacity of Spectrum Sharing System Using CSI and SSI under Received Power Constraint (Published Online). *Wireless Netw.*. **2018**. DOI: 10.1007/s11276-018-1666-7.

CHAPTER 17

Touch Screen Mobile Phones Form Analysis Using Kansei Engineering

VIVEK SHARMA* and KAMALPREET SANDHU

*Department of Product and Industrial Design,
Lovely Professional University, Phagwara, India*

*Corresponding author. E-mail: Sharma.v0516@gmail.com

ABSTRACT

Consumer perception about a product is linked with emotions evoked by it. The Kansei engineering is a methodology by which users' emotions or perception about a product can be identified more precisely and can be linked with product form elements such as shape, size, and color. In this research, a consumer preference–oriented design method of products' form perception is proposed touch screen mobile phone. The results of the study indicated that the design parameters of body shape and screen size received the highest importance weightage with 28.478 and 25.211 compared to others. Further, categories such as arc in top shape, line in bottom shape, radian in body shape, only touch in functional button, above 4.5 in. in screen size, above 1.95 in body ratio, and b/w 8.0 and 9.0 mm in thickness have the highest utility value compared to all others in their respective categories. An Internet of things (IoT)-based framework for this analysis is also developed for remote analysis.

17.1 INTRODUCTION

With rapid globalization competition, consumer product market has intensified to such a degree that it is harder to disguise one product from other in a particular segment based on traditional parameters such as price, technology, and usability.[1,2] Further, most of these products are more or less copy of each other and because of which consumer's buying behavior is becoming difficult to predict.[1] To gain the advantage over one competition is not only vital for success but also necessary for the survival of company.[11] Due to the increase in competition, success of a well-designed product depends not only on the functional requirements but also on product appearance for which the product form is vital.[1,3–5,9]

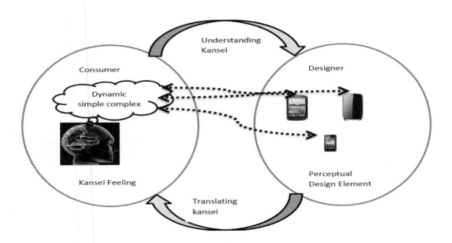

FIGURE 17.1 Kansei engineering mapping with product design elements.

Consumer perception about a product is linked with emotions evoked by it. The Kansei engineering is a methodology by which users' emotions or perception about a product can be identified more precisely and can be linked with product form elements such as shape, size, and color.[6–8] Kansei or evoked emotions about a product form can be identified using user response that can be identified by either psychophysiological signals such as electroencephalography (EEG), heart rate (HR), electromyography (EMG), and electrodermal activity (EDA) or users' questionnaire response on a semantic scale.[9,12] For example, it can be used to identify customer perception about corners or square shape of a product's edges by evaluating evoked images on semantic scales of words such as cool–dull, traditional–modern, and common–distinct using[10] methodology shown in Figure 17.1. Also, there is a lack of coordination in various stakeholders such as designers, engineers, and marketing salespersons in terms of product design ideal for consumer satisfaction. Hence, there is an urgent need to understand customer perception and develop the connecting tool between various stakeholders.

Since mobile phones are one of the widely used products in the world with a large competitive market, large brands like Samsung, Apple, and MI are top players in this market, which are producing more or less similar mobile phones (technology and price wise) in particular segment. In this research evaluation of consumer perception, firstly, mobile phone is carried out by linking quantitative relationship between customer's feelings (perception) and product form features, that is, mobile phones using the Kansei engineering using semantic-scale questionnaire. Secondly, IoT-based framework for this analysis, connecting various stakeholders (product design), is also developed.

17.2 MATERIALS AND METHODS

For evaluating Kansei (emotional appeal) about a product, two things are required: firstly, a list of product attributes and, secondly, emotional responses that are assessed by questionnaire method using semantic space. For listing product attributes (different form elements), the latest touch screen mobile phones that were in the same price range and with similar technology were chosen for this research with total of 20 phones. The images of all these chosen phones were downloaded from various ecommerce websites such as Flipkart, Amazon, and eBay. These images were preprocessed by scaling down, removing color, and brand information by using standard image processing software. The information of various form elements such as top shape, bottom shape, body shape, function key, screen size, body ratio, and thickness about all mobile phones was collected from manufacturer sites and tabulated. A variety of different form elements are listed in Table 17.1.

TABLE 17.1 Touch Screen Mobile Phones' Attributes.

S. no.	Type attribute	Variation of attributes			
1	A	Top shape	Arc	Curve	Line
2	B	Bottom shape	Arc	Curve	Line
3	C	Body shape	Arc	Curve	Line / Radian
4	D	Function key	Only touch	Touch + Rect round	Touch + Rect
5	E	Screen size	Below 4.0 in.	4.0–4.2 in.	4.2–4.5 in. / 4.5–5.0 in.
6	F	Body ratio	Below 1.92	B/w 1.92 and 1.95	Above 1.95
7	G	Thickness	Below 8	B/w 8 and 9 / B/w 9 and 9.5	B/w 9.5 and 11 / Above 11

The morphological analysis on all 20 mobile phones was performed and each mobile phone was listed as per different form element categories, for example, if mobile phone's top shape was arc, then it was listed as one in top shape. The morphological analyses of all mobile phones based on forms attributed are listed in Table 17.2.

For assessment of semantic space using the emotional words, that is, Kansei words, various words related to mobile phones were collected through various magazines, Internet searches, advertising vocabularies, etc. These words were screened, and relevant words for touch screen phones were analyzed and chosen by a group of experts. After this process, 11-word pairs of Kansei words were chosen and those are listed in Table 17.3. Web version of this experiment was created in OpenSesame.

The study design was approved by the ethical committee of Punjab Engineering College, Chandigarh. Participants for this study were recruited via flyer advertisement with free lunch coupons. In total, 138 university students (with 60% females) between the ages of 18–25 years volunteered for this study. All participants were informed about the study and methods used in the study and their informed consent were taken. All participants were instructed to perform experiment on OpenSesame and to rate each phone on every word on 5-point Likert scale. The responses, that is, word ratings from all participants were recorded in this survey.

TABLE 17.2 Morphological Chart of 20 Touch Screen Mobile Phone Based on Form Attributes.

Type attribute S. no.	Top shape (A)	Bottom shape (B)	Body shape (C)	Function button (D)	Screen size (E)	Body ratio (F)	Thickness (G)
1	1	1	4	1	2	1	3
2	1	1	4	1	4	3	4
3	1	1	4	1	1	1	5
4	1	1	2	1	4	2	3
5	2	2	3	1	2	1	4
6	1	1	4	1	4	2	3
7	1	1	1	3	3	1	2
8	1	1	4	3	1	1	4
9	1	1	4	3	4	1	3
10	1	1	4	1	4	2	2
11	1	2	4	3	2	2	4
12	1	1	4	2	3	1	2
13	3	3	3	2	2	1	3
14	1	1	1	2	1	1	5
15	1	1	2	2	1	2	5
16	3	3	3	2	2	3	4
17	1	1	4	2	3	3	2
18	3	3	3	2	4	3	1
19	1	2	1	2	1	1	5
20	2	2	1	2	2	3	5

TABLE 17.3 Kansei Words for Touch Screen Mobile Phones.

S. no.	Kansei words
1	Traditional–modern
2	Female–male
3	Complex–simple
4	Dull–cool
5	Delicate–rough
6	Babyish–mature
7	Boring–dynamic
8	Pointed–round
9	Heavy–thin
10	Common–distinct
11	Causal–business

TABLE 17.4 Classification of Image Sense Words.

	Component		
	1	2	3
Dull–cool	.947	.196	.017
Traditional–modern	.944	.157	.093
Boring–dynamic	.921	.118	−.069
Babyish–mature	.919	.273	.020
Delicate–rough	.761	.407	.053
Heavy–thin	.728	−.098	.340
Casual–business	.133	.941	.063
Common–distinct	.124	.937	.050
Complex–simple	.311	.576	.275
Female–male	.025	.113	.944
Pointed–round	.080	.131	.941

17.3 RESULTS

The results of survey are analyzed and separated word wise. To check intercorrelation between ratings of various words, factor analysis with varimax rotation using PCA algorithm was performed on all word ratings in SPSS software. The results of factor analyses are shown in

Table 17.4. The selected Kansei word can be classified into three groups or emotions, since strong intercorrelation ratings scored for these words, which is greater than $r > .5$ such as component no. 1 or emotion 1 includes dull–cool, traditional–modern, boring–dynamic, babyish–mature, delicate–rough, and heavy–thin. Component no. 2 includes casual–business, common–distinct, and complex–simple. Lastly, there are in between female–male and pointed–round, as per the results of factor analyses mentioned in Table 17.4. Also, in these three components, the rating score of Kansei word dull–cool has the highest intercorrelation with $r = .947$ for component no. 1, casual–business for component no. 2 with $r = .941$, and female–male for component no. 3 with $r = .944$, respectively. Hence, the ratings of Kansei words dull–cool, casual–business, and female–male are chosen for further analysis, some are highlighted in Table 17.4. Further, dull–cool has the highest correlation coefficient with $r = .947$ in all three emotions. Hence, it was selected for linking product form attributes to emotional states.

For linking mobile phone form attributes to emotional appeal, conjoint analysis was applied in this study. Conjoint analysis model uses regression analysis model with dummy variable with the relation described in the following equation.

$$P_h = b_0 + \sum_{i=1}^{I} b_i x_i + \in \qquad (17.1)$$

Further, using this method, the relationship between the utility of a product sample U and attribute-level a_{ij} is described in the following equation.

$$U(\text{dull} - \text{cool}) = b + \sum_{i=1}^{m} \sum_{j=1}^{ki} a_{ij} X_{ij} \qquad (17.2)$$

In the previous equation, U(dull-cool) is the total utility for the sample; a_{ij} is the jth level utility for the ith attribute; k is the number of attribute level; and m is the attribute number. If the jth level for the ith attribute exists, $X_{ij} = 1$. If not, $X = 0$. Through eq 17.2, the utility values of mobile phone form attributes were obtained and shown in Table 17.5.

The importance for each design parameter is obtained by using eq 17.3 as shown in Figure 17.2.

$$IMPi = 100 \frac{\text{Range } i}{\sum_{i}^{p} \text{Range } i} \qquad (17.3)$$

TABLE 17.5 Utility Value of Mobile Phone Form Attributes.

		Utility estimate
Top shape	Arc	.774
	Curve	.268
	Line	.505
Bottom shape	Arc	.294
	Curve	.125
	Line	.169
Body shape	Arc	1.795
	Curve	.520
	Line	1.213
	Radian	1.102
Function button	Only touch	.480
	Touch + Rect curve	.441
	Touch + Radian	.011
Screen size	Below 4 in.	1.269
	B/w 4.0 and 4.2 in.	.394
	B/w 4.2 and 4.5 in.	.651
	Above 4.5 in.	1.526
Body ratio	Below 1.92	.964
	B/w 1.92 and 1.95	.332
	Above 1.95	.632
Thickness	Below 8.0 mm	.063
	B/w 8.0 and 9.0 mm	.369
	B/w 9.0 and 9.5 mm	.064
	B/w 9.5 and 11.0 mm	.054
	Above 11.0 mm	.314
(Constant)		3.088

As shown in Figure 17.2, design parameters of body shape and screen size received the highest importance weightage with 28.478 and 25.211 compared to others. Further, categories such as arc in top shape, line in bottom shape, radian in body shape, only touch in functional button, above 4.5 in. screen size, above 1.95 in body ratio, and b/w 8.0 and 9.0 mm in thickness have the highest utility value compared to all others in their respective categories.

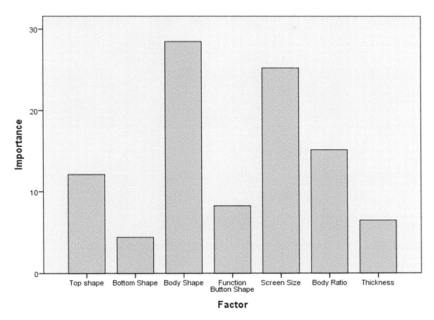

FIGURE 17.2 Importance of mobile phone form attributes.

17.4 DEVELOPMENT OF IOT-BASED FRAMEWORK

For sharing the result of customer perception obtained via the Kansei engineering, an IoT-based framework was developed as shown in Figure 17.3. It consists of survey making tool, addition of various importing images of products as described in table, addition of attribute list for each category of products as shown in Table 17.2, result analysis tool using methodology of conjoint analysis for obtaining utility scores for each product attributes, and, in the end, the sharing of result with all stakeholders such as designers, engineers, and marketing salepersons for the considerations of using IoT framework. This framework using block diagrams is shown in Figure 17.3.

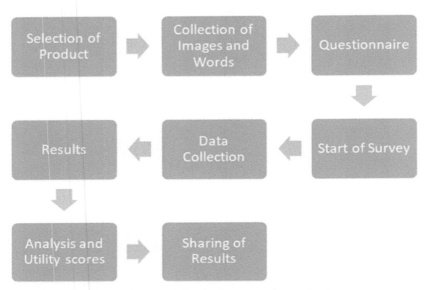

FIGURE 17.3 IoT-based framework for sharing perception evaluation.

17.5 DISCUSSION AND CONCLUSION

In this research, using the methodology of the Kansei engineering customer perception that is a subjective rating has been linked to product form attributes, that is, design parameters of touch screen phones. For this purpose, firstly, various emotional words linked to touch phones were collected from various websites, magazines, and newspapers. All collected words were screened by a group of experts and 11-word pairs were chosen. Similarly, 20 latest touch screen mobile phones were selected. Their specifications and pictures were collected from manufacturer websites. All pictures were normalized by scaling down, removing color, and brand information by using standard image processing software. The information of various form elements such as top shape, bottom shape, body shape, function key, screen size, body ratio, and thickness about all mobile phones was collected from manufacturer sites and tabulated. The questionnaires using images and words were created using OpenSesame software.

This questionnaire response from 138 university students (with 60% females) between the ages of 18–25 years was collected in the form of each word rating (5-point Likert scale for all 11-word pairs) from all participants. The results of the survey analyzed via factor analysis indicated strong intercorrelation between word rating that were classified into three groups or emotions such as component no. 1 or emotion 1 that includes dull–cool, traditional–modern, boring–dynamic, babyish–mature, delicate–rough, and heavy–thin. Component no. 2 includes casual–business, common–distinct, and complex–simple. Out of which, dull–cool scores the highest intercorrelation in component no. 1 with $r = .947$, casual–business in component no. 2 with $r = .941$, and female–male in component no. 3 with $r = .944$, respectively. Kansei word of dull–cool has the highest correlation coefficient with $r = .947$ in all three emotions. Hence, it was selected for linking product form attributes to emotional states using conjoint analysis. Using the utility obtained during conjoint analysis indicated that the design parameters of body shape and screen size received the highest importance weightage with 28.478 and 25.211 compared to others. Further, arc in top shape, line in bottom shape, radian in body shape, only touch in functional button, above 4.5 in. in screen size, 1.95 in body ratio, and b/w 8.0 and 9.0 mm in thickness category have the highest utility value compared to all others in their respective categories. These design parameters should be carefully considered using design process. With this method, designers can directly find out which features affect a certain image the most by comparing the important values of various design attribute features. In addition, for the alternative schemes, designers can easily find the best one that satisfies the image most by comparing the total utility values. Finally, an IoT-based framework for sharing the results about consumer perception is developed.

KEYWORDS

- **affective computing**
- **Kansei engineering**
- **mobile phone**
- **consumer perception**

REFERENCES

1. Bloch, P. H. Seeking the Ideal Form: Product Design and Consumer Response. *J. Market.* **1995**, *59* (3), 16–29.
2. Nigel, C. *Engineering Design Methods: Strategies for Product Design*; John Wiley & Sons Inc, 2000.
3. Hertenstein, J. H.; Platt, M. B.; Veryzer, R. W. The Impact of Industrial Design Effectiveness on Corporate Financial Performance. *J. Product Innov. Manage.* **2005**, *22* (1), 3–21.
4. Hsu, S. H.; Chuang, M. C.; Chang, C. C. A Semantic Differential Study of Designers' and Users' Product Form Perception. *Int. J. Ind. Ergonom.* **2000**, *25* (4), 375–391.
5. Lewalski, Z. M. *Product Esthetics: An Interpretation for Designers*; Design & Development Engineering PR, 1988.
6. Nagamachi, M. Kansei Engineering: A New Ergonomic Consumer-Oriented Technology for Product Development. *Int. J. Ind. Ergonom.* **1995**, *15* (1), 3–11.
7. Roy, R.; Goatman, M.; Khangura, K. User-Centric Design and Kansei Engineering. *CIRP J. Manuf. Sci. Technol.* **2009**, *1* (3). 172–178.
8. Jindo, T.; Hirasago, K. Application Studies to Car Interior of Kansei Engineering. *Int. J. Ind. Ergonom.* **1997**, *19* (2), 105–114.
9. Yamamoto, M.; Lambert, D. R. The Impact of Product Aesthetics on the Evaluation of Industrial Products. *J. Product Innov. Manage.* **1994**, *11* (4), 309–324.
10. Osgood, C. E.; Suci, G. J. *Factor Analysis of Meaning. Semantic Differential Technique—A Sourcebook*; Osgood, C. E., Snider, J. G., Eds.; 1969.
11. Sharma, V.; Prakash, N. R.; Kalra, P. User Sensory Oriented Product form Design Using Kansei Engineering and Its Methodology for Laptop Design. *IJSRSET* **2016**, *2* (1), 161–164.
12. Sharma, V.; Prakash, N. R.; Kalra, P. Audio-Video Emotional Response Mapping Based Upon Electrodermal Activity. *Biomed. Signal Process. Control*. **2019**, *47*, 324–333.

Index

A

AC parameters
  analog and RF performance
    capacitance of, 28
    transconductance (gm), 27
Area control error (ACE), 211
ATPG (Automatic Test Pattern Generation), 71
Automatic generation control (AGC), 210
  area to control, 212
  conventional control techniques
    integral controlling technique, 214
  HVDC (high voltage direct current) transmission, 212
  integrator, 217
  linear quadratic governor–based supervising, 214
  model predictive control (MPC), 213
  PID controller, 217
  pitch actuator, 217
  proportional integral (PI) controlling technique, 214
  proportional with integral derivative (PID), 214
  results and discussion
    BMFO2 algorithm, 218
    convergence curve and trial solutions, 219
    hybrid BMFO-SIG algorithm, 218
  simulation model, 216, 217
  SISO (Serial in Serial out), 213
  soft computing–based techniques
    artificial neural network, 215
    fuzzy logic, 215
    genetic algorithm (GA), 215
    particle swarm optimization, 215–216
  solar PV, 218
  turbine-governor, 218

B

BIST blocks, 71
Bits per pixel (BPP), 174
Bit-Swapping Complete Feedback Shift Register (BS-CFSR), 70
BIST architecture
  blocks, 75
  carry-lookahead (CL) adder, 74
  RC adder, 74
  methodology
    4-bit modified LFSR, 72
    Ex-OR gate, 72
    LFSR, series, 72
    n-bit with, 73
    ORA, 73–74
    pseudorandom nature, 72
  results and discussion
    4-bit ripple adder and CL adder, 75
    dynamic and total power consumption, 77
    fault-free response, 76
    Pass response, 76
    RTL, 75
Block discrete cosine transform (BDCT), 174
Blocking artifacts, 175
  BPP, 187, 192
    absolute error, 193
  existing method, 177
  fast DCT algorithm for, 179–181
  formation, 176
  images, absolute error, 192
  measurement, 176
  proposed detection method, 177–179
  proposed method with, 183, 186, 189–190
  proposed work, 175–176
  results and discussion, BPP, 182
  vertical and horizontal, 184–185, 191

C

Cadence's Encounter® RTL Compiler ripple carry (RC) 14.10, 71
Circuit under test (CUT), 70

Cognitive radio networks
    fading capacity, 272
        M-QAM scheme, 273
    instantaneous transmission power *versus* sensing metric, 274
    numerical results, 273
    SU adapts transmission power, 275
    SU transmission power, 274
    system model, 270
        proposed spectrum sharing scenario, 271
Complementary metal oxide semiconductor (CMOS)
    bootstrap driver
        layout of, 62
        limitation of, 60–61
        modified bootstrap driver, 61–62
        parameter, 65
        traditional driver, 59–60
    bootstrapping, concept, 52–53
        coupling ratio, 54
        current flow, 54
        dynamic CMOS logic, 57
        static CMOS logic, 58
        traditional bootstrap technique, 56–57
        transistor, 55

D

DC parameters
    threshold voltage
        drain-induced barrier lowering (DIBL), 27
        OFF-state, 26
        ON-state n-channel, 26
        slope of drain, 27
Delay-tolerant network (DTN) protocol, 197
Double-tail current dynamic latch comparator (DTDLC), 85–87

E

ECC (elliptic-curve cryptography), 100–101
Embedded system (ES)
    advantages, 144
    characteristics
        application and domain specific, 143
        connect input and output devices, 144
        distribution, 143–144
        harsh environment, operation, 143
        power concerns, 144
        reactive and real time, 143
        small size and weight, 144
        tightly constrained, 144
    disadvantages, 144
    hard real-time, 142
    large-scale, 143
    medium-scale, 143
    mobile, 142
    network, 142
    real-time, 141
    small-scale, 142–143
    soft real-time, 142
    stand-alone, 141

F

FEA (finite element analysis), 39
FPGA (Field Programmable Gate Array) sensors, 231–232
Fuzzy logic controller (FLC), 212

G

Gas sensor with micro dimensions
    Eigen frequency, 46
    meshed gas sensor 1, 44
    meshed gas sensor 2, 45
    resultant gas sensor 2, 45
    sensor consists, 44
    surface acoustic waves (SAW), 43
    Young's modulus E, 44
Genetic algorithm (GA), 211
Gradient descent, 211

H

HFSS (High Frequency Simulation Software), 137

I

Internet of thing (IoT), 223
    challenges and future expectations, 238
    communication in agriculture
        bluetooth-based moisture and temperature sensors, 234–235
        cellular communication, 234
        cloud computing, 236

Index

LORA, 235
SIGFOX, 235
Zigbee, 234
conventional farming technique
  fertilizers, 228–229
  irrigation, 227–228
  mapping and sampling of soil, 226–227
  monitoring forecasting and harvesting of yield, 229
  pest and crop disease, management, 229
smart agricultural technique, 224–225, 229
  acoustic sensors, 231
  airflow-type sensors, 232
  electrochemical sensors, 233
  FPGA (Field Programmable Gate Array) sensors, 231–232
  greenhouse farming, 230
  harvesting robots, 233
  hydroponic, 230
  optical sensors, 232
  optoelectronic-type sensors, 232
  phenotyping, 231
  sensors and their application, 233
  tractors, 233
  ultrasonic-type ranging sensors, 232
  vertical farming, 230
  wireless sensors, 231–233
soil moisture is compared to water deficit index (SWDI), 227
unmanned aerial vehicles
  crop, monitoring, 237
  detection of GAP, 237
  health assessment, 237
  irrigation, 237
  and plant counting, 237
  plant species, 238
  planting, 237
  soil and work site analysis, 236–237
  spray pesticides, 237
Internet of things (IoT), 2
  applications, 7
    blockchain technology, 15
    cloud computing, 15
    e-commerce, 11
    environment, 12
    e-waste, 12–13
    health-related fields, 10
    intelligent and smart transportation, 10
    intelligent metering, 10–11
    power grid, 11
    power supply, 12
    smart agriculture, 8
    smart cities, 11
    smart grid, 12
    smart home, 9
    smart industry, 8–9
    smart world, 9
    WSNs, 15
  architecture
    layered structure, 3–4
    major standard bodies, 4
    smart applications taxonomy, 5
  encryption and decryption algorithms
    AddRoundKey, 104
    advanced encryption standard, 102
    blowfish algorithm, 105
    cryptography, 99–100
    data encryption standard, 104–105
    data security features in, 98–99
    ECC (elliptic-curve cryptography), 100–101
    F-function, 102
    hybrid algorithm implementation, 108–109
    Inv. ShiftRows Step, 103
    Inv. SubByte, 103
    lightweight algorithms, 105–106
    MixColumns Step, 103–104
    RSA (Rivest–Shamir–Adleman) algorithm, 100
    secure force, 101–102
    steganography algorithms, 108
    triple data encryption standard, 105
    watermarking, 106–107
  issues and challenges
    data mining, 13
    integration, 14
    multiple network and locations, 14
    privacy, 13
    visualization, 13–14
  literature review, 5
  literature survey, 7

microstrip patch antenna, 136
  antenna fabrication, 138
  embedded system (ES), 139–144
  HFSS (High Frequency Simulation Software), 137
  PCB (Printed Circuit Board), 138
  RFID (Radio Frequency Identification), 137
  slotted antenna design, 137
  split-ring resonator (SRR), 138
  system, defined, 138–139
  wireless local area network (WLAN), 138
requirements, 6
video forensics in
  challenges, 126–129
  crime scene investigation, novel variables, 122–125
  ECHO gadget, 120–121
  electrocardiogram (ECG), 115
  forensics, 116–118
  intelligent transport system, 119
  tools, limitations in, 129
  typical smart home, 120
wireless sensor networks (WSNs), 8

K

Kansei engineering, 280
  IOT-based framework, development, 287
  framework for sharing perception evaluation, 288
  materials and methods, 281
    morphological analysis, 282
    OpenSesame, 283
    semantic space, 282
    touch screen mobile phones, 282
  results
    body shape and screen, 286
    Conjoint analysis model, 285
    PCA algorithm, 284
    U (dull-cool), 285
  smart phones, 281
    image sense words, 284
    morphological chart, 283–284
    utility value of, 286

L

Linear feedback shift register" (LFSR) techniques, 70–71
Load-frequency control (LFC), 210
Localized Encryption and Authentication Protocol (LEAP), 101
Low-power CMOS comparator
  comparison of
    double-tail current dynamic latch comparator (DTDLC), 85–87
    modified double-tail current dynamic latch comparator (MDTDLC), 88–89
    single-tail current dynamic latch comparator (STDLC), 82–85
    two-stage dynamic comparator without inverted lock (DTDLC-CLK), 90–92
  double-tail comparator, 81
  proximity sensor, 80
  result analysis
    CLK, 93
    comparator architectures, 92

M

MEMS (micro electro mechanical systems) devices
  advanced cantilevers with additional hole creation, 40
  capacitance plot, 43
  device design with result, 40–41
  electrostatic actuator with microbeams, 42
  FEA (finite element analysis), 39
  fundamentals of, 33
    advanced technique, 34
    beam deflection (z), 35
    Hook's law, 35
    implementation, 34
    mathematical equations, 34
    surface stress (δs), 35
  gas sensor with micro dimensions
    Eigen frequency, 46
    meshed gas sensor 1, 44
    meshed gas sensor 2, 45
    resultant gas sensor 2, 45

Index

sensor consists, 44
surface acoustic waves (SAW), 43
Young's modulus E, 44
Internet of Things (IoT)
  actuators and sensors, 37
  colossal advantages, 35
  connectivity, 37
  data processing, 38
  delivering and do something on information, 36
  gathering and forwarding information, 36
  sensors, 36–37
  user interface, 38
IoT potential markets, 47
microcantilevers, 38–39
SAW
  device 1, potential distribution, 46
  device 2, potential distribution, 47
Metal oxide semiconductor field effect transistors (MOSFETs), 19
  asymmetric gate MOSFET DGMOSFET, 24
  gate-all-around (GAA), 24–25
  asymmetric JLDG, 25
  biomedical applications, 29–30
  with biosensing cavity region, double-gate, 30
  cylindrical GAA, 25
  double-gate, 22
  FET-based memory design
    dynamic power dissipation, 28–29
  FinFET, 22
    3D and cross-sectional view, 23
    3-D schematic view of bottom spacer, 24
    3-D view of pi-gate, 23
  performance analysis
    AC parameters, 27–28
    DC parameters, 26–27
  SRAM cell
    power comparison of, 28
    7T FinFET, 29
  structures and dimensions
    bulk MOSFET, 20
    buried oxide region (BOX), 22–23
    multigate with SOI box region, 22
    SOI technology, 20–21
    triple-gate FET, 22
Microstrip patch antenna
  advantages and disadvantages, 153
  applications of, 153
  eligible effective dielectric constant, 149
  embedded system (ES), 139
    block diagram of, 140
    components of, 140
    hardware, 139
    real-time operating system, 139
    software, 139
    types of, 141–143
  feed line, 150
  internet of things (IoT), 144
    aperture coupling technique, 152
    applications, 147
    coaxial feed, 151
    design issues/challenges, 146–147
    embedded systems, role, 145–146
    microstrip feed line, 151
  patch thickness, 149
  radiation box calculation, 150
  requirement, 147
  softwares used in simulation, 154
  wireless standards with frequency bands, 148
Mobile ad hoc network (MANET), 197
  congestion factor, 199
  delay with and without optimization, 204
  energy consumption, 204
  flowchart of methodology, 200
  frame rate–based congestion, 199
  iHAR (improved Hotspot-based Adaptive Routing) method
    PDR *versus* time of proposed method, 202
  MATLAB simulator, 198
  proposed method, 199
  simulation results, 200
    iHAR (improved Hotspot-based Adaptive Routing) method, 201
    packet delivery ratio (PDR), 202–203
    static node, 205
  wireless sensor network (WSN), 199
Modified double-tail current dynamic latch comparator (MDTDLC), 88–89

N

National Institute of Standards and Technology (NIST) algorithm, 102

O

OFDM system model, 244
  BER plot, 255, 263
  carrier-to-interference ratio (CIR), 248
  CIR plot for various schemes, 262
  communication system, block diagram, 245
  FFT block processes, 246
  ICI cancellation schemes
    conjugate cancellation (CC) scheme, 250
    FFT, 251
    PRCC scheme, 250–251
    self-ICI cancellation scheme, 248–250
    symbol energies, 251–252
  inverse fast Fourier transform (IFFT) processor modulates, 245
  PCSC scheme (proposed scheme)
    AWGN channel, 254
    block diagram of, 253
    CFO, 252
    CPE, 252
    CPE-compensated discrete-time samples, 253
    data symbols, 254
  scatter plots of received signal, 262
  simulation results, 255–265
  weighting coefficients, 247
ORA (Output Response Analyzer), 71

P

Particle swarm optimization (PSO), 210
  flowchart, 217
PCB (Printed Circuit Board), 138

R

RFID (Radio Frequency Identification), 137
RSA (Rivest–Shamir–Adleman) algorithm, 100

S

Sensible vision, 158
  methods, 164
    analyze, 166
    available, 166
    database, 167
    flowchart, 165
    home, 165
    input, 166
    OpenCV, 167
    output, 166
    Phyton, 167
    processing information, 166
    start, 166
    talking, 167
    walking, 166
  Raspberry Pi, 162
    audio jack, 164
    Broadcom SoC, 162
    DSI connector, 164
    GPIO, 163–164
    power supply, 163
    processor, 162
    SD card, 163
  results, 167
    bottle detection program, 168
    home mode, 168
    human face expression, 169
    mobile detection program, 168
    objects detection program, 169
    talking mode, 169
    walk mode, 169
  systems
    audio jack, 161
    bluetooth, 161
    camera, 160–161
    3D sound device, 161–162
    deep learning, 159
    display, 160
    machine learning, 160
    microcontroller, 159
    microphone, 161
    neural network, 159–160
Single-tail current dynamic latch comparator (STDLC), 82–85
Slotted antenna design, 137

Index

Soil moisture is compared to water deficit index (SWDI), 227
Split-ring resonator (SRR), 138

T

Threshold voltage
　drain-induced barrier lowering (DIBL), 27
　OFF-state, 26
　ON-state n-channel, 26
　slope of drain, 27
Traditional bootstrap technique, 56–57
Traditional driver, 59–60
Transistor, 55
Triple data encryption standard, 105
Triple-gate FET, 22
Turbine-governor, 218
Two-stage dynamic comparator without inverted lock (DTDLC-CLK), 90–92

U

U (dull-cool), 285
Unmanned aerial vehicles
　crop, monitoring, 237
　detection of GAP, 237
　health assessment, 237
　irrigation, 237
　and plant counting, 237

plant species, 238
planting, 237
soil and work site analysis, 236–237
spray pesticides, 237

V

Video forensics in IoT
　challenges, 126–129
　crime scene investigation, novel variables, 122–125
　ECHO gadget, 120–121
　electrocardiogram (ECG), 115
　forensics, 116–118
　intelligent transport system, 119
　tools, limitations in, 129
　typical smart home, 120

W

Watermarking, 106–107
Wireless local area network (WLAN), 138

Y

Young's modulus E, 44

Z

Zigbee, 234